CONCEPTS IN ENTERPRISE

RESOURCE PLANNING

Third Edition

CONCEPTS IN ENTERPRISE

RESOURCE PLANNING

Third Edition

Ellen F. Monk
University of Delaware

Bret J. Wagner
Western Michigan University

COURSE TECHNOLOGY
CENGAGE Learning™

Australia • Brazil • Japan • Korea • Mexico • Singapore • Spain • United Kingdom • United States

COURSE TECHNOLOGY
CENGAGE Learning™

**Concepts in Enterprise Resource Planning,
  Third Edition**

by Ellen F. Monk and Bret J. Wagner

VP/Editorial Director: Jack W. Calhoun

Editor-in-Chief: Alex von Rosenberg

Senior Acquisitions Editor: Charles
  McCormick, Jr.

Product Manager: Tricia Coia, Kate Hennessy

Development Editor: Amanda Brodkin

Content Project Manager: Aimee Poirier

Manufacturing Coordinator: Justin Palmeiro

Marketing Manager: Bryant Chrzan

Editorial Assistant: Bryn Lathrop

Cover Designer: Laura Rickenbach

Cover Photo® 2007 Jupiterimages
  Corporation

Compositor: GEX Publishing Services

For product information and technology assistance, contact us at
Cengage Learning Academic Resource Center, 1-800-423-0563

For permission to use material from this text or product,
submit all requests online at **www.cengage.com/permissions**
Further permission questions can be emailed to
permissionrequest@cengage.com

ISBN-13: 978-1-4239-0179-2
ISBN-10: 1-4239-0179-7

Course Technology Cengage Learning
25 Thomson Place
Boston, Massachusetts, 02210
USA

Cengage Learning products are represented in Canada by Nelson
Education, Ltd.

For your lifelong learning solutions, visit **course.cengage.com**

Visit our corporate website at **www.cengage.com**

SAP is a registered trademark.

Printed in the United States of America
3 4 5 6 7 8 9 TW 12 11 10 09

In memory of our colleague Majdi Najm. His support and friendship are sorely missed.

BRIEF CONTENTS

TABLE OF CONTENTS

PREFACE

This is a book about Enterprise Resource Planning (ERP) systems; it's also about how a business works and how information systems fit into business operations. More specifically, it's about looking at the processes that make up a business enterprise and seeing how ERP software can improve the performance of these business processes. ERP software is complicated and expensive. Unless a company uses it to become more efficient and effective in delivering goods and services to its customers, the ERP system will only be a drain on company resources.

Prior to writing this book, our experience revealed that undergraduate business students don't always understand how businesses operate, and advanced undergraduate students and even many MBA students do not truly grasp the problems inherent in unintegrated systems. These students also do not comprehend business processes and how different functional areas must work together to achieve company goals. As a result, students do not understand how an information system should help business managers make decisions.

Consequently, we set out to write a book that does the following:

- Describes basic business functional areas and explains how they are related.

- Illustrates how unintegrated information systems fail to support business functions and business processes that cut across functional area boundaries.

- Demonstrates how integrated information systems can help a company prosper by improving business processes and by providing business managers with accurate, consistent, and current data.

We have found that our focus on business processes has been well received.

The Approach of This Book

A key feature of our book is the use of the fictitious Fitter Snacker Company, a manufacturer of nutritious snack bars, as an illustrative example throughout the book. We show how Fitter Snacker's somewhat primitive and unintegrated information systems cause operational problems. We intentionally made the systems' problems easy to understand, so the student could readily comprehend them. Potential solutions for solving integration problems are illustrated using SAP's ERP software.

The third edition of *Concepts in Enterprise Resource Management* reflects the current state of the ERP software market, while adding updated examples of how companies are using integrated systems to solve business problems and achieve greater success. The book has eight chapters:

- Chapter 1 explains the purposes for, and information systems requirements of, **main business functional areas**—Marketing and Sales, Supply Chain Management, Accounting and Finance, and Human Resources. This chapter also describes how a business process cuts across the activities within business functional areas and why managers need to think about making business processes work.

- Chapter 2 provides a short history of business computing and the **developments that led to today's ERP systems**. Chapter 2 concludes with an overview of ERP issues and an introduction to the SAP ERP software.

- Chapter 3 describes the **Marketing and Sales functional area**, and it highlights the problems that arise with unintegrated information systems. To make concepts easy to understand, the Fitter Snacker running example is introduced. After explaining FS's problems with its unintegrated systems, we show how ERP can avoid these problems. SAP ERP screens are used to illustrate the concepts. Because using ERP can naturally lead a company into ever-broadening integration, a discussion of customer-relationship management (CRM) concludes the chapter.

- Chapter 4 describes how ERP systems support **Supply Chain Management**—the coordinated activities of all the organizations involved in converting raw materials into consumer products on the retail shelf. As in Chapter 3, the problems caused by Fitter Snacker's unintegrated information system are explored, followed by a discussion of how ERP software could help solve these problems.

- Chapter 5 describes **Accounting and ERP systems.** This chapter clearly distinguishes between financial accounting (FI) and managerial accounting (CO) issues. Included is an overview of the Enron collapse and the resulting Sarbanes-Oxley act along with the act's impact on information systems, specifically management controls and audit capabilities.

- Chapter 6 covers **Human Resource Management**. While the Human Resource software is the least integrated component of all ERP systems, it includes numerous processes that are critical to a company's success, including strategic issues like succession planning.

- Chapter 7 covers **Process Modeling, Process Improvement, and ERP Implementation**. The chapter first presents flowcharting basics using a minimum number of symbols, followed by the highly structured EPC process model. Implementation issues conclude the chapter. We believe that process improvement, not large-scale implementation, should be the focus in an introductory ERP course.

- Chapter 8 covers ERP and electronic commerce. Because this is a broad and rapidly changing area, we have chosen to provide this chapter as an introduction, rather than an exhaustive treatment of the subjects. This chapter provides an overview of topics such as electronic commerce and application service providers, SAP NetWeaver, and the emerging technology of RFIDs.

How Can You Use This Book?

This third edition continues our goal of keeping the text at an introductory level. The book can be used in a number of ways:

- The book, or selected chapters, could be used for a three-week ERP treatment in undergraduate Management Information Systems, Accounting Information Systems, or Operations Management courses.

- Similarly, the book or selected chapters could be used in MBA courses, such as foundation Information Systems or Operations Management courses. Although the concepts presented here are basic, the astute instructor can build on them with more sophisticated material to challenge the advanced MBA student. Many of the exercises in the book require research for their solution, and the MBA student could do these in some depth.

- The book could serve as an introductory text in a course devoted wholly to ERP. It would provide the student with a basis in how ERP systems help companies to integrate different business functions. The instructor might use Chapter 8 as the starting point for teaching the higher-level strategic implications of ERP and related topics. The instructor can pursue these and related topics using his or her own resources, such as case studies and current articles.

- Because of the focus on fundamental business issues and business processes, the book can also be used in a sophomore-level Introduction to Business course.

Except for a computer literacy course, we assume no particular educational or business background. Chapters 1 and 2 lay out most of the needed business and computing groundwork, and the rest of the chapters build on that base.

Features of This Text

To bring ERP concepts to life (and down to earth!) this book uses sales, manufacturing, purchasing, human resources, and accounting examples for the Fitter Snacker company. Thus, the student can see problems, not just at an abstract level, but within the context of a company's operations. We believe that this approach makes business problems and the role ERP can play in solving them easier to understand.

The book's exercises have the student analyze aspects of Fitter Snacker's information systems in various ways. The exercises vary in their difficulty; some can be solved in a straightforward way, and others require some research. Not all exercises need to be assigned. This gives the instructor flexibility in choosing which concepts to emphasize and how to assess students' knowledge. Some exercises explore FS's problems, and some ask the student to go beyond what is taught in the book and to research a subject. A solution might require the student to generate a spreadsheet, perform calculations, document higher-level reasoning, present the results of research in writing, or participate in a debate.

The book includes an additional element designed to bring ERP concepts to life: Another Look features, which are short, detailed case studies that focus on problems faced by real-world companies. Some of these cases include interviews with information systems managers who share their experiences with ERP.

We have illustrated ERP concepts and applications by showing how SAP ERP would handle the problems discussed in the book. Screen shots of key SAP ERP tools are shown throughout to illustrate ERP concepts. Many of the book's exercises ask the student to think about how a problem would be addressed using ERP software.

Instructor Materials

The following supplemental materials are available when this book is used in a classroom setting. All of the teaching tools available with this book are provided to the instructor on a single CD-ROM. Most can also be found online at www.course.com. Instructor materials are password-protected.

- **Electronic Instructor's Manual**—The Instructor's Manual assists in class preparation by providing suggestions and strategies for teaching the text, chapter outlines, technical notes, quick quizzes, discussion topics, and key terms.

- **Solutions**—Answers to end-of-chapter questions and exercises are provided.

- **Sample syllabi**—The sample syllabi and course outlines are provided as a foundation to begin planning and organizing your course.

- **ExamView Test Bank**—ExamView allows instructors to create and administer printed, computer (LAN-based), and Internet exams. The Test Bank includes hundreds of questions that correspond to the topics covered in this text, enabling students to generate detailed study guides that include page references for further review. The computer-based and Internet testing components allow students to take exams at their computers, and also save the instructor time by grading each exam automatically. The Test Bank is also available in Blackboard and WebCT versions posted online at www.course.com.

- **PowerPoint Presentations**—Microsoft PowerPoint slides for each chapter are included as a teaching aid for classroom presentation, to make available to students on the network for chapter review, or to be printed for classroom distribution. Instructors can add their own slides for additional topics they introduce to the class.

- **Distance learning**—Course Technology is proud to present online test banks in WebCT and Blackboard to provide the most complete and dynamic learning experience possible. Instructors are encouraged to make the most of the course, both online and offline. For more information on how to access the online test bank, contact your local Course Technology sales representative.

- **Figure Files**—Figure and table files from each chapter are provided for your use in the classroom.

- **Hands-on SAP exercises**—Exercises are available for member institutions through the SAP University Alliance. These exercises use a database that was built for the fictitious Fitter Snacker company.

ACKNOWLEDGMENTS

Our thanks go out to our development editor, Amanda Brodkin, who learned how to work with authors who occasionally use too much jargon and have trouble meeting deadlines, and provided the critical eye that we needed to make our writing into what we imagined it was. We are grateful for the support and guidance of the entire MIS team at Course Technology, particularly managing editor Tricia Coia and production editor Aimee Poirier. We would not have been able to continue on our journey to understand ERP systems without the continued support of SAP America through its University Alliance program. We appreciate the efforts of Amelia Maurizio, Heather Czech Matthews and Doug Peebles. We also thank our reviewers Sam Gill; San Francisco State University and Cindy Joy Marselis, Temple University, for insightful comments that pointed to needed improvements. In addition, we thank our interviewees, Maureen Sullivan, Linda Somers, Ellen Lepine, John Wheeler, and Pat Ryan with his colleagues from DuPont, as well as Phil Straniero and Don Scott, for their time and frankness. And finally, we thank our students, whose honesty and desire to learn have inspired us.

CHAPTER **1**

BUSINESS FUNCTIONS AND BUSINESS PROCESSES

LEARNING OBJECTIVES

After completing this chapter, you will be able to:

- Name the main functional areas of operation used in business.
- Differentiate a business process from a business function.
- Identify the kinds of data that each main functional area produces.
- Identify the kinds of data that each main functional area needs.
- Define integrated information systems and explain why they are important.

INTRODUCTION

Enterprise Resource Planning (ERP) programs are core software used by companies to coordinate information in every area of the business. ERP (pronounced "E-R-P") programs help to manage company-wide business processes, using a common database and shared management reporting tools. A **business process** is a collection of activities that takes one or more kinds of input and creates an output, such as a report or forecast, that is of value to the customer. ERP software supports the efficient operation of business processes by integrating throughout a business tasks related to sales, marketing, manufacturing, logistics, accounting, and staffing. In later chapters, you will learn how successful businesspeople use ERP programs to improve how work is done within a company. This chapter provides a background for learning about ERP software.

FUNCTIONAL AREAS AND BUSINESS PROCESSES

To understand ERP, you must first understand how a business works. Let's begin by looking at a business's areas of operation. These areas, called **functional areas of operation**, are broad categories of business activities.

Functional Areas of Operation

Most companies have four main functional areas of operation: **Marketing and Sales (M/S)**, **Supply Chain Management (SCM)**, **Accounting and Finance (A/F)**, and **Human Resources (HR)**. Each area comprises a variety of narrower **business functions**, which are activities specific to that functional area of operation. For example, the business functions of each area for some companies are shown in Figure 1-1.

Functional area of operation	Marketing and Sales	Supply Chain Management	Accounting and Finance	Human Resources
Business functions	Marketing of a product	Purchasing goods and raw materials	Financial accounting of payments from customers and to suppliers	Recruiting and hiring
	Taking sales orders	Receiving goods and raw materials	Cost allocation and control	Training
	Customer support	Transportation and logistics	Planning and budgeting	Payroll
	Customer relationship management	Scheduling production runs	Cash-flow management	Benefits
	Sales forecasting	Manufacturing goods		Government compliance
	Advertising	Plant maintenance		

FIGURE 1-1   Examples of functional areas of operation and their business functions

Historically, businesses have had organizational structures that separated the functional areas, and business schools have been similarly organized, so each functional area has been taught as a separate course. In a company separating functional areas in this way,

Marketing and Sales might be completely isolated from Supply Chain Management, even though M/S sells what SCM procures and produces. Thus, you might conclude that what happens in one functional area is not closely related to what happens in others. As you will learn in this chapter, however, functional areas are interdependent, each requiring data from the others. The better a company can integrate the activities of each functional area, the more successful it will be in today's highly competitive environment. Integration also contributes to improvements in communication and workflow. Each area's information system depends on data from those of other functional areas. An **information system (IS)** includes the computers, people, procedures, and software that store, organize, and deliver information. This chapter illustrates the need for information sharing between functional areas and the effects on the business if this information is not integrated. You will also see examples of typical business processes and how these processes routinely cross functional areas.

Business Processes

Recently, managers have begun to think in terms of business processes rather than business functions. Recall that a business process is a collection of activities that takes one or more kinds of input and creates an output that is of value to the customer. The customer for a business process can be the traditional external customer (the person who buys the finished product), or it may be an internal customer (such as a colleague in another department). For example, what is sold through M/S is linked to what is procured and produced by SCM. This concept is illustrated in Figure 1-2.

Input	Functional area responsible for input	Process	Output
Request to purchase computer	Marketing and Sales	Sales order	Order is generated
Financial help for purchase	Accounting and Finance	Arranging financing in-house	Customer finances through the computer company
Technical support	Marketing and Sales	24-hour help line available	Customer's technical query is resolved
Fulfillment of order	Supply Chain Management	Shipping and delivery	Customer receives computer

FIGURE 1-2 Sample business processes related to the sale of a personal computer

Thinking in terms of business processes helps managers to look at their organization from the customer's perspective. For example, suppose that a customer wants to purchase a new computer. She wants information about the company's products so she can select a computer and various peripherals. She wants to place her order quickly and easily, and perhaps arrange for financing through the company. She expects quick delivery of a correctly configured, working computer, and she wants 24-hour customer support for any problems. The customer is not concerned about how the computer was marketed, how its components were purchased, or how it was built, or how the delivery truck will find the best route to her house. The customer wants the satisfaction of having a working computer at a reasonable price.

Businesses must always consider the customer's viewpoint in any transaction. What is the difference between a business function and a business process from the customer's point of view? Suppose the customer's computer is damaged during shipment. Because only one functional area is involved in accepting the damaged item, receipt of the return is a *business function* and is handled by the customer relationship management function of Marketing and Sales. Because several functional areas are involved in repair and return of the computer, the handling of the repair is a *business process*. Thus, the customer is dealing with many of the company's functional areas in her process of buying and obtaining a computer.

A successful customer interaction is one in which the customer (either internal or external) is not required to interact with each business function involved in the process. Successful business managers view their business operations from the perspective of a satisfied customer.

For the computer company to provide customer satisfaction, it must make sure that its functional areas of operation are integrated. For example, computer technology changes rapidly, and the hardware the computer company sells changes frequently. In order to provide customers with accurate information, people performing the sales function must have up-to-date information about computer configurations; otherwise, a customer might order a computer that the company's manufacturing plant no longer produces. People performing the manufacturing function need to get the details of a customer's computer configuration quickly and accurately from the employees performing the sales function, so the right computer can be manufactured and shipped on time to the customer. If the customer is financing the computer through the computer company, then people performing the sales order function must gather information about the customer and process it quickly, so financing can be approved in time to support shipping the computer.

Sharing data effectively and efficiently between and within functional areas leads to more efficient business processes. Information systems can be designed so that functional areas share data. These systems are called **integrated information systems**. Working through this textbook will help you understand the benefits of integrated information systems and the problems that can occur when information systems are not integrated. Figure 1-3 illustrates the process view of business operations.

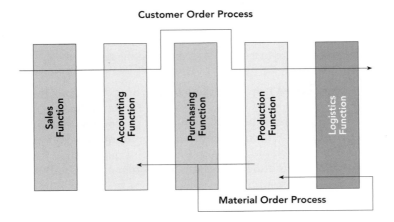

FIGURE 1-3 A process view of business

Businesses take inputs (resources) in the form of material, people, and equipment, and transform these inputs into goods and services for customers. Managing these inputs and the business processes effectively requires accurate and up-to-date information. For example, the sales function takes a customer's order, and the production function schedules the manufacturing of the product. Logistics employees schedule and carry out the delivery of the product. If raw materials are needed to make the product, production prompts purchasing to arrange for their purchase and delivery. In that case, logistics will receive the material, verify its condition to accounting so that the vendor can be paid, and deliver the goods to production. Throughout, accounting keeps appropriate transaction records.

ANOTHER LOOK

Integrated Information Systems

Integration of information is essential for company efficiency. Although people in organizations are often bombarded with too much information, it can still be challenging to get the correct information to the department that needs it. The appliance giant Whirlpool Corporation has faced just that challenge. Whirlpool is committed to Enterprise Resource Planning systems and in 2000 began a huge implementation of an SAP ERP system (you will learn about SAP in Chapter 2).

Managing price increases is a particular challenge. With rising oil prices and increased raw material costs, Whirlpool needs to be able to look at the business as a whole—in other words, globally. Its integrated SAP system helps it to do just that.

Whirlpool corporate vice president and chief information officer Esat Sezer explains that, regarding raw material price increases, "We had to have the capability to see product by product, category by category, country by country, day to day, the impact of material costs, logistics costs, and the impact into our (profit) margins." With reference to the supply chain, in particular, Whirlpool has updated and fixed an unintegrated system that used to consist of spreadsheets and manual procedures. Now, with its integrated supply chain, demand from a trade partner or customer is integrated into production planning. "We can look into production plans and see if this item for this date in this quantity is for this customer," says Sezer.

Some of Whirlpool's midsized trading partners could not connect directly to Whirlpool's order entry system, and instead were ordering either by phone or fax, which was extremely inefficient. With the SAP integrated system, and a new online order system, these partners now can place orders over the Web. These improvements have translated into a savings of 80 percent of the cost of taking the order.

continued

Many other large corporations have similar integration stories. The DuPont Corporation is committed to SAP for integrating business units. For example, DuPont Fluorochemicals couldn't do division-wide capacity planning, nor could its customers order over the Web, until DuPont integrated its systems with one ERP system.

Other companies need to integrate information from acquisitions. For example, Air France is integrating its inventory and order systems after acquiring KLM.

When Colgate-Palmolive needed global reporting and global analysis along with more accurate and more timely cash positions, it looked to ERP systems to enable those capabilities.

Question:

1. Choose an industry in which you would enjoy working, and pick a company in that industry. Assume this company is lacking an integrated information system. Write a memo to the CEO explaining the benefits of integrating the systems in the company. Refer to the Functional Area Information Systems section of the chapter to help you compose this memo.

FUNCTIONAL AREAS AND BUSINESS PROCESSES OF A VERY SMALL BUSINESS

As an introduction, we will look at the way business processes involve more than one functional area, using a very small business as an example—a fictitious lemonade stand that you own. We will examine the business processes of the lemonade stand and see why coordination of the functional areas helps achieve efficient and effective business processes. You will see the role that information plays in this coordination and how integration of the information system improves your business.

Even though one person can run a lemonade stand, the operation of the business requires a number of processes. Coordinating the activities within different functional areas requires accurate and timely information.

Marketing and Sales

The functions of Marketing and Sales include developing products, determining pricing, promoting products to customers, and taking customers' orders. M/S also helps to create a sales forecast to ensure the successful operation of the lemonade stand.

For the most part, this is a cash business and does not require formal recordkeeping, but you still need to keep track of your customers so that you can send flyers or occasional thank you notes to repeat customers. Thus, your records must not only show the amount of sales, but also identify repeat customers.

Product development can be done informally in such a simple business; you gather information about who buys which kind of lemonade and note what customers say about each product. You also analyze historical sales records to spot trends that are not obvious. Deciding whether to sell a product also depends on how much it costs to produce the product. For example, some customers might be asking for a sugar-free lemonade. To determine whether the new lemonade could be profitably produced and sold, you could analyze data from SCM, including production information (such as mixing container size, time required to mix) and materials management data (cost of lemons and sweetener).

Even though you run a cash business, good repeat customers are allowed to charge purchases—up to a point. Thus, your records must show how much each customer owes and his or her available credit. It is very important that the data be available and accurate at the time of a customer's credit request. Since Accounting and Finance records must be accessed as a part of the selling process, the accounting function has a critical role to play in the sales process.

Supply Chain Management

The functions within Supply Chain Management include making the lemonade (manufacturing/production) and buying raw materials (purchasing). Production is planned so that, as much as possible, lemonade is available when needed, without excess production of lemonade that must be liquidated. This planning requires sales forecasts from the M/S functional area. **Sales forecasts** are analyses that attempt to predict the future sales of a product. A forecast's accuracy will be improved if it is based on historical sales figures (for example, factors such as hot weather and nearby yard sales will impact the forecast). Thus, forecasts from M/S play an important role in the production planning process.

Production plans are also used to develop requirements for raw materials (bottled spring water, fresh lemons, artificial sweetener, and raw sugar) and packaging (cups, straws, and napkins). You must generate raw material and packaging orders from these requirements. If the forecasts are accurate, you will not lose sales because of material shortages, nor will you have excessive inventory that might spoil.

SCM and M/S must choose a recipe for each lemonade product sold. The standard recipe is a key input for deciding how much to order of each raw material, which is a purchasing function. Access to this recipe is also necessary for keeping good manufacturing records, allowing managers within the SCM functional area (working with those in A/F) to compare how much it actually costs to make a glass of lemonade, versus how much the recipe *should* have cost.

Accounting and Finance

Functions within Accounting and Finance include recording raw data about transactions (including sales), raw material purchases, payroll, and receipt of cash from customers. **Raw data** are simply numbers collected from those operations, without any manipulation, calculation, or arrangement for presentation. Those data are then summarized in meaningful ways to determine the profitability of the lemonade stand and to support decision making.

Note that data from Accounting and Finance are used by Marketing and Sales as well as by Supply Chain Management. The sales records are an important component of the sales forecast, which is used in making staffing decisions and in production planning. The records from accounts receivable, which you use to determine whether to grant credit to a particular customer, are also used to monitor the overall credit-granting policy of the lemonade stand. You want to be sure that you have enough cash on hand to purchase raw materials, as well as to finance purchasing new equipment, such as a lemon juicer.

Human Resources

Even a simple business needs employees to support the M/S and SCM functional areas, which means that the business must recruit, train, evaluate, and compensate employees. These are the functions of Human Resources.

At the lemonade stand, the number of employees and the timing of hiring depend on the level of lemonade sales. HR uses sales forecasts developed by the individual departments to plan personnel needs. A part-time helper might be needed at forecasted peak hours or days. How much should a part-time helper be paid? That depends on prevailing job market conditions, and it is HR's job to monitor those conditions.

Would increased sales justify hiring a part-time worker at the prevailing wage? Or, should you think about acquiring more automated ways of making lemonade, so that a person working alone could run the stand? Resolving these questions requires input from SCM and A/F.

The lemonade stand, while a simple business, has many of the processes needed in larger organizations, and these processes involve activities in more than one functional area. In fact, it is impossible to discuss the processes in one functional area without discussing the links to other functional areas—connections that invariably require the sharing of data. Systems that are integrated using ERP software provide the data sharing that is necessary between functional areas.

FUNCTIONAL AREA INFORMATION SYSTEMS

The lemonade stand provides simple examples of business processes and the functional activities required to support them. Next we will describe potential inputs and outputs for each functional area (refer back to Figure 1-2 to review inputs and outputs related to the sale of a personal computer). Note the kinds of data needed by each area and how people use the data. Also note that the information systems maintain relationships between all functional areas and processes.

Marketing and Sales

The Marketing and Sales area needs information from all other functional areas to do its job. See Figure 1-4. Customers communicate their orders to M/S in person or by telephone, e-mail, fax, the Web, and so on. In the case of Web-based systems, customer and order data should be stored automatically in the information system; otherwise, data must be stored manually, by a person typing data into a keyboard or point-of-sale system, or using a bar code reader or other device. Sales orders must be passed to SCM for planning purposes and to A/F for billing. Sales order data are also valuable for analyzing sales trends for business decision making. For example, M/S management might use a report showing the trend of a product's sales to evaluate marketing efforts and to determine strategies for the sales force.

M/S also has a role in determining product prices, which requires an understanding of the market competition and the costs of manufacturing the product. Pricing might be determined based on a product's unit cost, plus some percentage markup. For example, if a product costs $5 per unit to make, and management wants a 40 percent markup, the selling price must be $7 per unit. Where does the per-unit cost come from? Determining the cost of manufacturing a product requires information from Accounting and Finance, which, in turn, relies on Supply Chain Management data.

People are a valuable asset to the firm, and M/S needs to interact with Human Resources to exchange information on hiring needs, legal requirements, and other matters. For example, when M/S has an opening for a junior salesperson, Human Resources will do the advertising for the job vacancy. Human Resources also conveys information about travel reimbursement to salespeople on the road. To summarize, inputs for M/S include:

- Customer data
- Order data
- Sales trend data
- Per-unit cost
- Travel expense company policy

Outputs for M/S include:

- Sales strategies
- Product pricing
- Employment needs

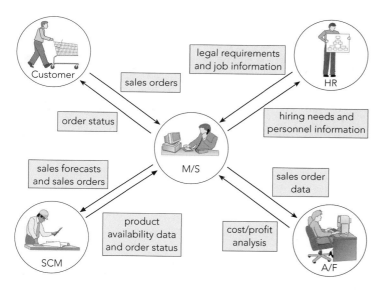

FIGURE 1-4 The Marketing and Sales functional area exchanges data with customers and with the Human Resources, Accounting and Finance, and Supply Chain Management functional areas

Supply Chain Management

Supply Chain Management also needs information from the various functional areas, as shown in Figure 1-5. Manufacturing firms develop production plans of varying length and detail, such as long-range, medium-range, and short-range plans. Each plan deals with different functional areas of the business. Examples of this planning might include expanding manufacturing capacity, hiring new workers, paying extra overtime for existing workers, and taking sales forecasts to plan manufacturing runs.

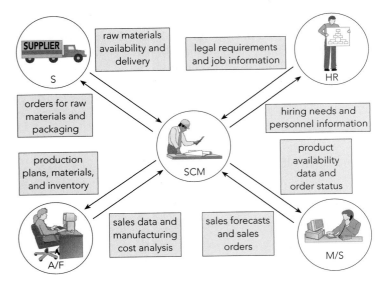

FIGURE 1-5 The Supply Chain Management functional area exchanges data with suppliers and with the Human Resources, Marketing and Sales, and Accounting and Finance functional areas

Production plans are based on information about product sales (actual and projected) that comes from Marketing and Sales. The purchasing function bases its orders of raw materials on production plans, expected shipments, delivery lead times, and existing inventory levels. With accurate data about required production levels, raw material and packaging can be ordered as needed, and inventory levels can be kept low, saving money. On the other hand, if data are inaccurate or not current, manufacturing may run out of raw material or packaging; such a shortfall is called a **stockout**. Shortages of this type can shut down production and cause the company to miss delivery dates. To avoid stockouts, management might carry extra raw material and packaging, known as **safety stock**, which can result in an overinvestment in inventory. If certain time-sensitive goods are held too long, they can spoil and will need to be destroyed, rather than sold for profit. The accuracy of the forecast determines the amount of safety stock that is required to reduce the risk of a stockout to an acceptable level. The less accurate the forecast, the more safety stock is required. Accurate forecasting and production planning can reduce the need for extra inventory and manufacturing capacity.

Supply Chain Management records can provide the data needed by Accounting and Finance to determine how much of each resource (materials, labor, supplies, and overhead) was used to make completed products in inventory.

Supply Chain Management data can support the M/S function by providing information about what has been produced and shipped. For example, some computer manufacturers, such as Gateway, have automated systems that call customers to notify them that their computer order has been shipped. Shipping companies, such as UPS and FedEx, provide shipment-tracking information on the Internet. By entering a tracking number, the customer can see each step of the shipping process by noting where the package's bar code was scanned. Thus, accurate and timely production information can support the sales process and increase customer satisfaction.

For long-range planning, management might want to see monthly reports showing sales and production figures. The data for such reports must come from the production and inventory data.

Supply Chain Management also interacts in some ways with Human Resources. SCM passes hiring information to HR, and HR informs SCM of the company's layoff and recall policy, which might pertain to workers in the plant.

To summarize, inputs for SCM include:

- Product sales data
- Production plans
- Inventory levels
- Layoff and recall company policy

Outputs for SCM include:

- Raw material orders
- Packaging orders
- Resource expenditure data
- Production and inventory reports
- Hiring information

Accounting and Finance

Accounting and Finance needs information from all the other functional areas to complete its jobs accurately, as depicted in Figure 1-6. A/F personnel record the company's transactions in the books of account. For example, they record accounts receivable when sales are made and cash receipts when customers send in payments. In addition, they record accounts payable when raw materials are purchased and cash outflows when they pay for materials. Finally, A/F personnel summarize the transaction data to prepare reports about the company's financial position and profitability.

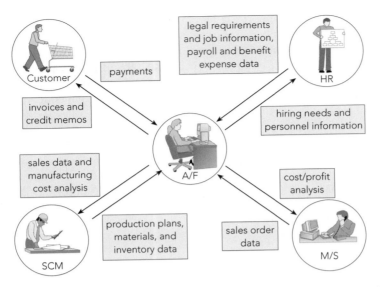

FIGURE 1-6 The Accounting and Finance functional area exchanges data with customers and with the Human Resources, Marketing and Sales, and Supply Chain Management functional areas

People in other functional areas provide data to A/F: M/S provides sales data, SCM provides production and inventory data, and HR provides payroll and benefit expense data. The accuracy and timeliness of A/F data depend on the accuracy and timeliness of the data from the other functional areas.

M/S personnel require data from A/F to evaluate customer credit. If an order will cause the customer to exceed his or her credit limit, M/S should see that the customer's accounts receivable balance (the amount owed to the company) is too high and hold new orders until the customer's balance is lowered. If A/F is slow to record sales, the accounts receivable balances will be inaccurate, and M/S might approve credit for customers who have already exceeded their credit limits and who might never pay off their accounts. If A/F does not record customers' payments promptly, the company could deny credit to customers who actually owe less than their credit limit, potentially damaging the company's relationship with those customers.

To summarize, inputs for A/F include:

- Payments from customers
- Accounts receivable data
- Accounts payable data
- Sales data
- Production and inventory data
- Payroll and expense data

Outputs for A/F include:

- Payments to suppliers
- Financial reports
- Customer credit data

Human Resources

Like the other functional areas, HR also needs information from the other departments to do its job accurately. See Figure 1-7. Tasks related to employee hiring, benefits, training, and government compliance are all the responsibilities of a Human Resources department. People in HR need accurate forecasts of personnel needs from all functional units. HR also needs to know what skills are needed to perform a particular job and how much the company can afford to pay employees. These data also come from all functional units.

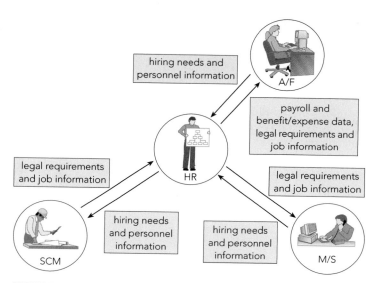

FIGURE 1-7 The Human Resources functional area exchanges data with the Accounting and Finance, Marketing and Sales, and Supply Chain Management functional areas

State and federal laws require companies to observe many governmental regulations in recruiting, training, compensating, promoting, and terminating employees—and these regulations must be observed company-wide. Usually, it is also HR's responsibility to ensure that employees receive training in a timely manner, and that they get certified (and recertified) in key skills, such as materials handling and equipment operation. HR must also administer wages, salaries, raises and bonuses. For these and other reasons, corporate HR needs timely and accurate data from other areas.

HR must create accurate and timely data for management use. For example, HR should maintain a database of skills required to do particular jobs and the prevailing pay rates. When the company evaluates employees' performance and compensation, analysis of these data can help to prevent the loss of valued employees because of low pay.

To summarize, inputs for HR include:

- Personnel forecasts
- Skills data

Outputs for HR include:

- Regulation compliance
- Employee training and certification
- Skills database
- Employee evaluation and compensation

As shown in Figures 1-4 through 1-7, a significant amount of data is maintained by and shared among the functional areas. The timeliness and accuracy of these data are critical to each area's success and to the company's ability to make a profit and generate future growth. ERP software allows all the functional areas to share a common database so that accurate, real-time information is available. In the next chapter, we will trace the evolution of data management systems that led to ERP.

Chapter Summary

- Companies that make and sell products have business processes that involve these basic functional areas: Marketing and Sales, Supply Chain Management, Accounting and Finance, and Human Resources. They perform these functions:

  - Marketing and Sales sets product prices, promotes products through advertising and marketing, takes customer orders, supports customers, and creates sales forecasts.

  - Supply Chain Management develops production plans, orders raw materials from suppliers, receives the raw material into the facility, manufactures products, maintains facilities, and ships products to customers.

  - Accounting and Finance performs financial accounting to provide summaries of operational data in managerial reports, and also is responsible for tasks such as controlling accounts, planning and budgeting, and cash-flow management.

  - Human Resources recruits, hires, trains, and compensates employees, ensures compliance with government regulations, and oversees the evaluation of employees.

- A functional area is served by an information system. Information systems capture, process, and store data to provide information needed for decision making.

- Employees working in one functional area need data from employees in other functional areas. Ideally, functional area information systems should be integrated, so shared data are accurate and timely.

- Today, business managers try to think in terms of business processes that integrate the functional areas, thus promoting efficiency and competitiveness. An important aspect of this integration is the need to share information between functions and functional areas. ERP software provides this capability by means of a single common database.

Key Terms

Accounting and Finance (A/F)	Integrated information system
Business function	Marketing and Sales (M/S)
Business process	Raw data
Enterprise Resource Planning (ERP)	Safety stock
Functional areas of operation	Sales forecast
Human Resources (HR)	Stockout
Information system (IS)	Supply Chain Management (SCM)

Exercises

1. Distinguish between a business function and a business process. Describe how a business process cuts across functional lines in an organization. Why do managers organize their teams in terms of business processes instead of functional departments? What benefits do you see from this new method of organization?

2. Reproduce Figure 1-1 for the lemonade stand business example. Add a one-sentence description for each function as it relates to the lemonade stand.

3. Assume you run an Internet business with a couple of friends from college. Your company sells tickets to concerts and sporting events. Describe all the flows of information between Marketing and Sales and Accounting and Finance.

4. Using the Internet, research your state's regulations for waiters and waitresses, such as minimum age of employment. Why is it important that Human Resources communicate this information to the hiring department?

5. Think of the last time you bought a pair of shoes. How does the process of buying those shoes cut across the store's various functional lines? What information from your receipt would need to be available to the business functions? Which business functions would need that information?

For Further Study and Research

Albright, Evan. "Whirlpool: Actionable Analytics." *SAP NetWeaver magazine* 02, no. 03 (Summer 2006).

Anthes, Gary H. "Case Study: Supply Chain Whirl." *Computerworld*, September 6, 2005. http://www.computerworld.com.au/index.php/id;1550407298;fp;2;fpid;1239068928.

IBM Corporation. Case Study, Whirlpool: "Whirlpool's B2B trading portal cuts per-order costs significantly." 2000. http://www-07.ibm.com/hk/e-business/case_studies/manufacturing/whirlpool.html.

Pang, Albert. IDC case study: "Air France Soars to New Heights with Upgraded my SAP ERP System: Meeting Its Commitment to Excellence." July 2006. http://www.sap.com/solutions/business-suite/erp/pdf/CCS_Air_France.pdf.

SAP Business Community Webcast. 2005. http://www.sap.com/community/pub/events/2005_12_01_cp/index.epx.

Sullivan, Laurie. "ERPzilla," *Information Week*, July 11, 2005.

CHAPTER 2

THE DEVELOPMENT OF ENTERPRISE RESOURCE PLANNING SYSTEMS

LEARNING OBJECTIVES

After completing this chapter, you will be able to:

- Identify the factors that led to the development of Enterprise Resource Planning (ERP) systems.
- Describe the distinguishing modular characteristics of ERP software.
- Discuss the pros and cons of implementing an ERP system.
- Summarize ongoing developments in ERP.

INTRODUCTION

In today's competitive business environment, companies try to provide customers with goods and services faster and less expensively than their competition. How do they do that? Often, the key is to have efficient, integrated information systems. Increasing the efficiency of information systems results in more efficient management of business processes. When companies have efficient business processes, they can be more competitive in the marketplace.

An Enterprise Resource Planning (ERP) system can help integrate a company's operations by acting as a company-wide computing environment that includes a database that is shared by all functional areas. Such software can deliver consistent data across all business functions in real time. Real time refers to data and processes that are always current.

This chapter will help you to understand how and why ERP systems came into being and what the future might hold for business information systems. The chapter follows this sequence:

- Review of the evolution of information systems and related causes for the recent development of ERP systems

- Discussion of the few ERP software vendors that dominate the market. The current industry leader, German software maker SAP AG, and its industry-leading software product, SAP ERP, are discussed as an example of an ERP system.

- Review of factors influencing a company's decision on whether to purchase an ERP system

- Description of ERP's benefits

- Overview of frequently asked questions related to ERP systems

- Discussion of the future of ERP software and its impact on the Internet and Web services

THE EVOLUTION OF INFORMATION SYSTEMS

Until recently, most companies had unintegrated information systems that supported only the activities of individual business functional areas. Thus, a company would have a Marketing information system, a Production information system, and so on, each with its own hardware, software, and methods of processing data and information. This configuration of information systems is known as **silos** because each department has its own stack, or silo, of information that is unconnected to the next silo. Silos are also known as stovepipes.

Such unintegrated systems might work well within individual functional areas, but to achieve its goals, a company must share data among all the functional areas. When a company's information systems are not integrated, costly inefficiencies can result. For example, suppose two functional areas have separate, unintegrated information systems. To share data, a clerk in one functional area needs to print out data from another functional area and then type the information into her area's information system. Not only does this data input take twice the time, it also significantly increases the chance for data entry errors. Alternatively, the process might be automated by having one information system write data to a file to be read by another information system. This would reduce the probability of errors, but it could only be done periodically (usually overnight or on a weekend), to minimize the disruption to normal business transactions. Because of the time lag in updating the system, the transferred data would rarely be up to date. In addition, data can be defined differently in different data systems, such as calling products by different part numbers in different systems. This variance can create further problems in timely and accurate information sharing between functional areas.

It seems obvious today that a business should have integrated software to manage all functional areas. An integrated ERP system, however, is an incredibly complex hardware and software system that was not feasible until the 1990s. Current ERP systems evolved as a result of three things (1) the advancement of hardware and software technology (computing power, memory, and communications) needed to support the system, (2) the development of a vision of integrated information systems, and (3) the reengineering of companies to shift from a functional focus to a business process focus.

Computer Hardware and Software Development

Computer hardware and software developed rapidly in the 1960s and 1970s. The first practical business computers were the mainframe computers of the 1960s. Although these computers began to change the way business was conducted, they were not powerful enough to provide integrated, real-time data for business decision making. Over time, computers got faster, smaller, and cheaper, leading up to today's proliferation of mobile devices. The rapid development of computer hardware capabilities has been accurately described by Moore's Law. In 1965, Intel employee Gordon Moore observed that the number of transistors that could be built into a computer chip doubled every 18 months (see Figure 2-1). This meant that the capabilities of computer hardware were doubling every 18 months.

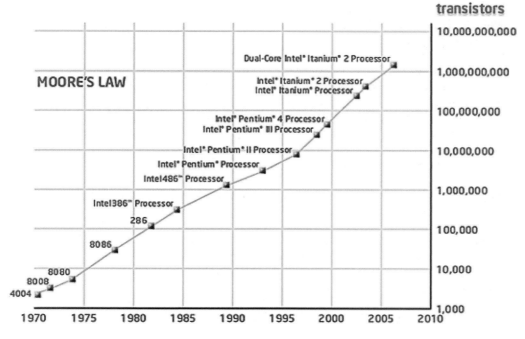

FIGURE 2-1 The actual increase in transistors on a chip approximates Moore's Law (Courtesy of Intel Corporation)

During this time, computer software was also advancing to take advantage of the increasing capabilities of computer hardware. In the 1970s, relational database software was developed, providing businesses with the ability to store, retrieve, and analyze large volumes of data. Spreadsheet software, a fundamental business tool today, became popular in the 1980s. With spreadsheets, managers could perform complex business analyses without having to rely on a computer programmer to develop custom programs.

The computer hardware and software developments of the 1960s and 1970s paved the way for the development of ERP systems.

Early Attempts to Share Resources

As PCs gained popularity in business in the 1980s, managers became aware that important business information was being stored on individual PCs, but that there was no easy way to share the information electronically. Users needed a way to share costly peripheral equipment (such as printers and hard disks, which in the early 1980s were fairly expensive) and, more importantly, data.

By the mid-1980s, telecommunications developments allowed users to share data and peripherals on local networks. Usually, these networks were groups of computers connected to one another within a single physical location. This meant that workers could download data from a central computer to their desktop PCs and work with the data at their desks.

This central computer–local computer arrangement is now called a **client-server architecture**. Servers (central computers) became more powerful and less expensive and provided scalability. **Scalability** means that when a piece of equipment's capacity is exhausted, its capacity can be increased by adding new hardware. In the case of a client-server network, the ability to add servers makes the network scalable—thus extending the life of the hardware investment.

By the end of the 1980s, much of the hardware needed to support the development of ERP systems was in place: fast computers, networked access, and advanced database technology. Recall from Chapter 1 that ERP programs help to manage company-wide business processes using a common database. This common database holds a very large amount of data. The technology to hold that data in an organized fashion, and to retrieve data easily, is the **database management system**, known as a **DBMS**. By the mid-1980s, the DBMS required to manage the development of complex ERP software existed. The final element required for the development of ERP software was understanding and acceptance from the business community. Businesspeople did not yet recognize the benefits of integrated information systems, nor were they willing to commit the resources to develop ERP software.

The Manufacturing Roots of ERP

The concept of an integrated information system took shape on the factory floor. Manufacturing software developed during the 1960s and 1970s, evolving from simple inventory-tracking systems to **material requirements planning (MRP)** software. MRP software allowed a plant manager to plan production and raw materials requirements by working backward from the sales forecast, the prediction of future sales. Thus, the manager first looked at Marketing and Sales' forecast of demand (what the customer wants), then looked at the production schedule needed to meet that demand, calculated the raw materials needed to meet production, and finally, projected raw materials purchase orders to suppliers. For a company with many products, raw materials, and shared production resources, this kind of projection was impossible without a computer to keep track of various inputs.

The basic functions of MRP could be handled by mainframe computers. **Electronic data interchange (EDI)**, the direct computer-to-computer exchange of standard business documents, allowed companies to handle the purchasing process electronically, avoiding the cost and delays resulting from paper purchase order and invoice systems. The functional area now known as Supply Chain Management (SCM) began with the sharing of long-range production schedules between manufacturers and their suppliers.

Management's Impetus to Adopt ERP

The hard economic times of the late 1980s and early 1990s caused many companies to downsize and reorganize. These company overhauls were a stimulus to ERP development. Companies needed to find some way to avoid the following kind of situation, which they had tolerated for a long time.

Imagine you are the CEO of Styling, a clothing manufacturing company. Styling is profitable and is keeping pace with the competition, but your IS is unintegrated and inefficient (as are the systems of your competitors). You've learned to live with this kind of inefficiency: Your Marketing and Sales department creates a time-consuming paper trail for negotiating and making a sale. To schedule factory production, however, your Manufacturing manager needs accurate, timely information about actual and projected sales orders from the M/S manager. Without such information, the Manufacturing manager must guess which products to produce—and how many of them to produce. To keep goods moving through the production line, the manager often does guess. Sometimes the guess overestimates demand for some garments, and sometimes it underestimates demand.

Overproduction of a certain garment might mean your company is stuck with garments for which there is no market, or you might face a diminishing market due to style changes or seasonal demand. When you store the garments, waiting for a buyer, you incur warehouse expense. On the other hand, underproduction of a certain garment might result in garments not being ready for delivery when a salesperson promised, leading to unhappy customers and canceled orders. If you try to catch up on orders, you'll have to pay factory workers overtime, or resort to the extra expense of rapid-delivery shipments.

The management of large companies decided they could no longer afford the type of inefficiencies illustrated by the Styling example—inefficiencies caused by the functional model of business organization. This model, illustrated in Figure 2-2, had deep roots in U.S. business, starting with the General Motors organizational model developed by Alfred P. Sloan in the 1930s. The functional business model illustrates the concept of silos of information, which limit the exchange of information between the lower operating levels. Instead, the exchange of information between operating groups is handled by top management, which might not be knowledgeable about the functional area.

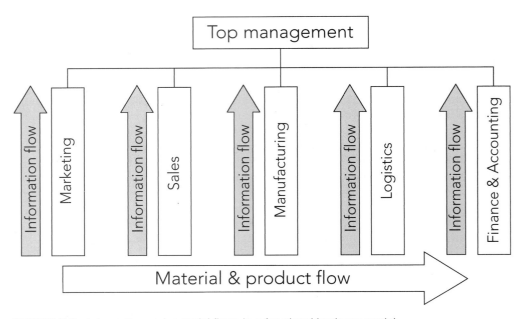

FIGURE 2-2 Information and material flows in a functional business model

The functional model was very useful for decades, and was successful in the United States, where there was limited competition and where flexibility and rapid decision making were not requirements for success. In the quickly changing markets of the 1990s, however, the functional model led to top-heavy and overstaffed organizations incapable of reacting quickly to change. The time was right to view a business as a set of cross-functional processes, as illustrated in Figure 2-3. In this organizational model, the functional business model, with its separate silos of information, is gone. Now information flows between the operating levels without top management's involvement.

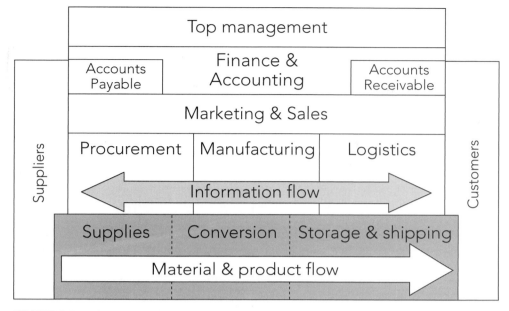

FIGURE 2-3 Information and material flows in a process business model

In a process-oriented company, the flow of information and management activity is "horizontal" across functions, in line with the flow of materials and products. This horizontal flow promotes flexibility and rapid decision making. Michael Hammer's 1993 landmark book, *Reengineering the Corporation: A Manifesto for Business Revolution*, stimulated managers to see the importance of managing business processes. Books like Hammer's, along with the difficult economic times of the late 1980s, led to a climate in which managers began to view ERP software as a solution to business problems.

In recent years, further impetus for adopting ERP systems has come from compliance with the Sarbanes-Oxley Act of 2002, a federal law passed in response to the accounting fraud discovered at Enron and WorldCom. The law requires companies to substantiate internal controls on all information. Sarbanes-Oxley is covered in Chapter 5. In the next section, you will learn about the development of the first ERP software. SAP was the first company to develop software for ERP systems and is the current market leader in ERP software sales. According to some estimates, SAP is used to complete 50 percent of the world's business transactions. As of 2007, SAP had 33,000 customers and seeks to triple that number by 2010.

ERP SOFTWARE EMERGES: SAP AND R/3

In 1972, five former IBM systems analysts in Mannheim, Germany—Dietmar Hopp, Claus Wellenreuther, Hasso Plattner, Klaus Tschira, and Hans-Werner Hector—formed *Systemanalyse und Programmentwicklung* (Systems Analysis and Program Development, or SAP, pronounced "S-A-P"). The computer industry of the time was quite different from that of today. IBM controlled the computer market with its 360 mainframe computer, which had only 512K of main memory. In this mainframe computer environment, the SAP founders recognized that all companies developing computer software faced the same basic business problems, and each developed unique, but similar, solutions for their needs in payroll processing, accounting, materials management, and other functional areas of business. SAP's goal was to develop a standard software product that could be configured to meet the needs of each company. SAP's concept from the beginning was to set standards in information technology, according to founder Dietmar Hopp. In addition, the founders wanted data available in real time, and they wanted users to work on a computer screen, not with voluminous printed output. These goals were lofty and forward-looking for 1972, and it took almost 20 years to achieve them.

SAP Begins Developing Software Modules

Before leaving IBM, Plattner and Hopp had worked on an order-processing system for the German chemical company ICI. The order-processing system was so successful that ICI managers also wanted a materials and logistics management system—a system for handling the purchase, receiving, and storage of materials—that could be integrated into the new order-processing system. In the course of their work for ICI, Plattner and Hopp had already developed the idea of modular software development. Software **modules** are individual programs that can be purchased, installed, and run separately, but that all extract data from the common database. Instead of giving Plattner and Hopp the new ICI project, IBM made them traveling software experts, removing them from hands-on development duties. Plattner and Hopp approached Claus Wellenreuther, an expert in financial accounting who had just left IBM, about forming their own company.

When Plattner, Hopp, and Wellenreuther established SAP on April 1, 1972, they could not afford to purchase their own computer. Their first contract, with ICI, to develop the follow-on materials and logistics management system, included access to ICI's mainframe computer at night—a practice they repeated with other clients until they acquired their first computer in 1980. At ICI, the SAP founders developed their first software package, variously called System R, System RF (for real-time financial accounting), and R/1.

To keep up with the ongoing development of mainframe computer technology, in 1978 SAP began developing a more integrated version of its software products, called the R/2 system. In 1982, after four years of development, SAP released its R/2 mainframe ERP software package.

Sales grew rapidly in the 1980s, and SAP extended its software's capabilities and expanded into international markets. This was no small task, because the software had to be able to accommodate different languages, currencies, accounting practices, and tax laws.

By 1988, SAP had established subsidiaries in numerous foreign countries, established a joint venture with consulting company Arthur Andersen, and sold its 1,000th system. SAP also became SAP AG, a publicly traded company.

SAP R/3

In 1988, SAP realized the potential of client-server hardware architecture and began development of its **R/3** system to take advantage of client-server technology. The first version of SAP R/3 was released in 1992. Each subsequent release of the SAP R/3 software contained new features and capabilities. The client-server architecture used by SAP allowed R/3 to run on a variety of computer platforms, including UNIX and Windows NT. The SAP R/3 system was also designed using an open architecture approach. In **open architecture**, third-party software companies are encouraged to develop add-on software products that can be integrated with existing software. The open architecture also makes it easy for companies to integrate their hardware products, such as bar code scanners, personal digital assistants (PDAs), cell phones, and global information systems with the SAP system.

New Directions in ERP

In the late 1990s, the Year 2000, or Y2K, problem motivated many companies to move to ERP systems. As it became clear that the date turnover from December 31, 1999 to January 1, 2000 would wreak havoc on some information systems, companies searched for ways to consolidate data, and ERP systems provided one solution.

The Y2K problem originated from programming shortcuts made by programmers in the preceding decades. With memory and storage space a small fraction of what it is today, early programmers developed software that used as few computer resources as possible. To save memory, programmers in the 1970s and 1980s typically wrote programs that only used two digits to identify a year. For example, if an invoice was posted on October 29, 1975, the programmer could just store the date as 10/29/75, rather than 10/29/1975. While this may not seem like a big storage savings, with millions of transactions needing to be manipulated, it adds up. These programmers never imagined that software written in the 1970s would still be running major companies and financial institutions in 1999. These old systems were known as **legacy systems**. Many companies were faced with a choice: pay programmers millions of dollars to correct the Y2K problem in their old, limited software—or invest in an ERP system that would not only solve the Y2K problem, but potentially provide better management of their business processes as well. Thus, the Y2K problem led to a dramatic increase in business for ERP vendors in the late 1990s. However, the rapid growth of the 1990s was followed by an ERP slump starting in 1999. By 1999, companies were in the final stages of either an ERP implementation or modification of their existing software. Many companies that had not yet decided to move to a Y2K-compliant ERP system waited until after the new millennium to upgrade their information systems.

By 2000, SAP AG had 22,000 employees in 50 countries and 10 million users at 30,000 installations around the world. By that time, SAP had competition in the ERP market, namely from Oracle and PeopleSoft (PeopleSoft expanded its offerings through the acquisition of ERP software vendor JD Edwards in 2003). In late 2004, Oracle succeeded in its bid to take over PeopleSoft.

PeopleSoft

PeopleSoft was founded by David Duffield, a former IBM employee who, like SAP's founders, faced opposition from IBM for his ideas. PeopleSoft started with software for human resources and payroll accounting, and achieved considerable success, even with companies that already were using SAP for accounting and production. PeopleSoft's success caused SAP to make significant modifications to its Human Resources module. PeopleSoft strengthened its offerings in the supply chain area with its acquisition of JD Edwards. Today, PeopleSoft, under Oracle, is a popular software choice for managing human resources and financial activities at universities.

Oracle

Oracle is SAP's biggest competitor. Oracle began in 1977 as Software Development Laboratories (SDL). Its founders, Larry Ellison, Bob Miner, and Ed Oates, won a contract from the Central Intelligence Agency (CIA) to develop a system, called Oracle, to manage large volumes of data and extract information quickly. Although the Oracle project was canceled before a successful product was developed, the three founders of SDL saw the commercial potential of a relational database system. In 1979, SDL became Relational Software, Inc. and released its first commercial database product. The company changed its name again, to Oracle, and in 1986 released the client-server Oracle relational database. The company continued to improve its database product, and in 1988 released Oracle Financials, a set of financial applications. The financial applications suite of modules included Oracle Financials, Oracle Supply Chain Management, Oracle Manufacturing, Oracle Project Systems, Oracle Human Resources, and Oracle Market Management. Oracle Financials was the beginning of what would become Oracle's ERP product.

The concepts of an enterprise resource planning system are similar for large vendors, such as SAP and Oracle, and for the many smaller vendors of ERP software. Because of SAP's leadership in the ERP industry, this textbook focuses primarily on SAP's ERP software products as an example of ERP. Keep in mind that most other ERP software vendors provide similar functionality, with some having strengths in certain areas.

SAP ERP

SAP ERP software (previous versions were known as R/3, and later, mySAP ERP) has changed over the years due to product evolution and for marketing purposes. The latest versions of ERP systems by SAP and other companies allow all business areas to access the same database, as shown in Figure 2-4, eliminating redundant data and communications lags. Perhaps most importantly, the system allows data to be entered once, and then used throughout the organization. In information systems, errors most frequently occur where human beings interact with the system. ERP systems ensure that data are entered only once, where they are most likely to be accurate. For example, with access to real-time stock data, a salesperson taking an order can confirm the availability of the desired material. When the salesperson enters the sales order into the system, the order data are immediately available to Production, so Manufacturing can update production plans, and Materials Management can plan the delivery of the order. If the sales order data are entered correctly by the salesperson, then SCM personnel are working with the same, correct data. The same sales data are also available to Accounting for preparing invoices.

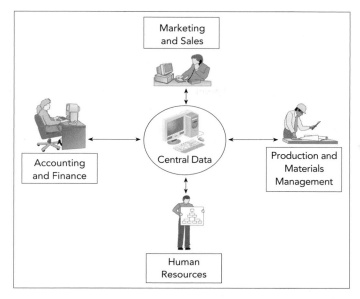

FIGURE 2-4   Data flow within an integrated information system

Earlier in this chapter, you learned how software modules work. Figure 2-5 on the next page shows the major functional modules in the current SAP ERP system, also known as SAP ECC 6.0 (Enterprise Central Component 6.0), and depicts how the modules provide integration.

The basic functions of each of the modules are as follows:

- The **Sales and Distribution (SD) module** records sales orders and scheduled deliveries. Information about the customer (pricing, how and where to ship products, how the customer is to be billed, and so on) is maintained and accessed from this module.
- The **Materials Management (MM) module** manages the acquisition of raw materials from suppliers (purchasing) and the subsequent handling of raw materials inventory, from storage to work-in-progress goods to shipping of finished goods to the customer.
- The **Production Planning (PP) module** maintains production information. Here production is planned and scheduled, and actual production activities are recorded.
- The **Quality Management (QM) module** plans and records quality control activities, such as product inspections and material certifications.

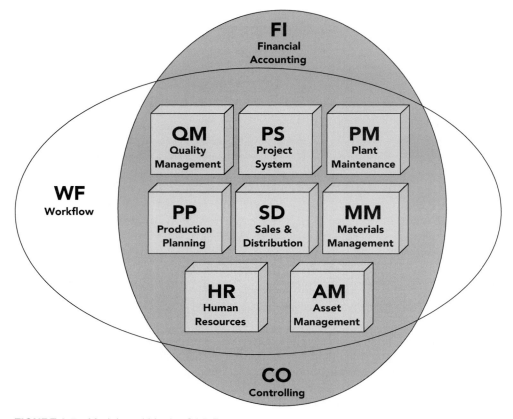

FIGURE 2-5   Modules within the SAP ERP integrated information systems environment
(Courtesy of SAP AG)

- The **Plant Maintenance (PM) module** manages maintenance resources and planning for preventive maintenance of plant machinery, to minimize equipment breakdowns.
- The **Asset Management (AM) module** helps the company to manage fixed-asset purchases (plant and machinery) and related depreciation.
- The **Human Resources (HR) module** facilitates employee recruiting, hiring, and training. This module also includes payroll and benefits.
- The **Project System (PS) module** allows the planning for and control over new R&D, construction, and marketing projects. This module allows for costs to be collected against a project, and it is frequently used to manage the implementation of the SAP ERP system. PS manages build-to-order items, which are low-volume, highly complex products such as ships and aircrafts.

Two financial modules, FI and CO, are shown in Figure 2-5 as encompassing the modules described above. That is because nearly every activity in the company has an impact on the financial position of the company.

- The **Financial Accounting (FI) module** records transactions in the general ledger accounts. This module generates financial statements for external reporting purposes.
- The **Controlling (CO) module** serves internal management purposes, assigning manufacturing costs to products and to cost centers, so that the profitability of the company's activities can be analyzed. The CO module supports managerial decision making.
- The **Workflow (WF) module** is not a module that automates a specific business function. Rather, it is a set of tools that can be used to automate any of the activities in SAP ERP. It can perform task-flow analysis and prompt employees (by e-mail) if they need to take action. Workflow is ideal for business processes that are not daily activities, but that occur frequently enough to be worth the effort to implement workflow, such as preparing customer invoices.

To summarize: ERP integrates business functional areas with one another. Before ERP, each functional area operated independently, using its own information systems and ways of recording transactions. ERP software also makes management reporting and decision making faster and more uniform throughout an organization. In addition, ERP promotes thinking about corporate goals, as opposed to thinking only about the goals of a single department or functional area. When top management is queried on the reasons for implementing ERP systems, the overriding answer is *control*. With the capability to see integrated data on their entire company's operation, managers use ERP systems for the control they provide, allowing them to set those corporate goals correctly.

SAP ERP Software Implementation

A truly integrated information system entails integrating all functional areas, but for various reasons, not all companies that use SAP use all of the SAP ERP modules. For example, a company without factories wouldn't choose the manufacturing-related modules. Another company might consider its HR department's operations to be so separate from its other operations that it would not integrate its HR functional area. Another company might believe that its internally developed production and logistics software gives it a competitive advantage, and so it would implement the SAP ERP Financial Accounting and Human Resources modules, but integrate its internally developed production and logistics system into the SAP ERP system.

Generally, a company's level of data integration is highest when the company uses one vendor to supply all of its modules. When a company uses modules from different vendors, additional software programming must be done to get the modules to work together. Frequently, companies integrate different systems using batch data transfer processes that are performed periodically. In this case, the company no longer has accurate data available in real-time across the enterprise. Thus, a company must be sure the decision to use multiple vendors or to maintain a legacy system is based on sound business analysis, not on a resistance to change. Software upgrades of nonintegrated systems are made more problematic because further work must be done to get software from different vendors to interact. SAP's NetWeaver development platform (discussed in Chapter 8) eases the integration of SAP ERP with other software products.

Any large software implementation is challenging—and ERP systems are no exception. There are countless examples of large implementations failing, and it's easy to understand why. Many different departments are involved, as are many users of the system, programmers, systems analysts, and other personnel. Without top management commitment, large projects are doomed to fail. More implementation issues are discussed in Chapter 7.

After a company chooses its major modules, it must make an incredibly large number of decisions on how to configure the system. These configuration options allow the company to customize the modules it has chosen to fit the company's needs. For example, in the FI module, a business might need to define limits on the dollar value of business transactions that an employee can process. This is an important consideration in minimizing the risk of fraud and abuse.

Tolerance Groups

In configuring the SAP system, the company can define **tolerance groups**, which are specific ranges that define these transaction limits. An example of a tolerance group is shown in Figure 2-6. As part of the configuration process, a company can define any number of tolerance groups with a range of limits and can then assign employees to these tolerance groups.

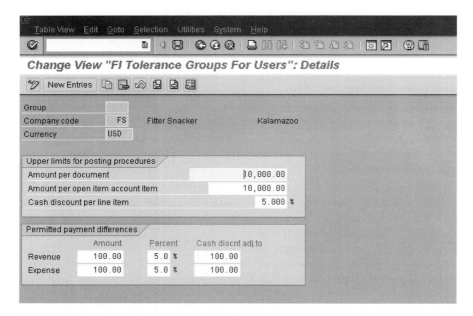

FIGURE 2-6   A customization example: tolerance groups to set transaction limits

While SAP has defined the tolerance group methodology as its method for placing limits on an employee, configuration allows the company the flexibility to further tailor this methodology. Let's assume an outdoor clothing company places an order in July for 1,000 ski jackets. The company receives only 995 in the shipment from the manufacturer, which arrives in September. This delivery, although it is short five ski jackets, is close enough to the original order that it is accepted as complete. The difference of the five ski jackets represents the tolerance. By defining the tolerance group to accept a variance of a small percentage of the shipment, the company has determined that it is not worth pursuing the five extra ski jackets. Tolerance could indicate a shortage, as in this example, or an overabundance in an order. Thus, an order of 1,005 ski jackets would also be within the tolerance. Tolerance groups should be defined and documented, in part to deal with fraud issues. The clothing company should know the reason for a short order: is it because the order is within the tolerance range, or is it because the workers on the loading dock stole five ski jackets?

Features of SAP ERP

Not only was SAP ERP the first software that could deliver real-time ERP integration, it has other features worthy of note. Its most significant characteristics are usability by large companies, high cost, automation of data updates, and applicability of best practices, as described below.

The original SAP R/3 system targeted very large companies. Prior to the development of ERP systems, it was assumed that these giants could never have integrated systems because of the sheer amount of computing power required to integrate them. Increased computing speeds, however, meant that large companies in a variety of industries, including manufacturing, gas and oil, airlines, and consulting, could have integrated information systems.

Acquiring an SAP ERP software system is very expensive. In addition to the cost of the software, many companies find they must buy new hardware to accommodate such powerful programs. For a Fortune 500 company, software, hardware, and consulting costs can easily exceed $100 million. Large companies can also spend $50 million to $100 million on upgrades. Full implementation of all modules can take years. In fact, most companies view ERP implementations as an ongoing process, not a one-off project. As implementations are completed in one area of a company, other areas may begin an implementation or upgrade a previous implementation.

The modular design of SAP ERP is based on business processes, such as sales order handling, materials requirement handling, and employee recruiting. When data are entered into the system, data in all related files in the central database are automatically updated. No human input is required to make the changes.

Before the development of SAP ERP, IS people felt that software should be designed to reflect a business's practices. Companies sought vendors to write software to fit their business processes. As SAP accumulated experience developing information systems, however, the company began to develop models of how certain industries' business processes should be managed. Thus, SAP ERP's design incorporates **best practices**, which means that SAP's software designers choose the best, most efficient ways in which business processes *should* be handled. If a user's business practices don't follow one of the best practices incorporated in the SAP ERP design, then the business must redesign its practices so it can use the software, sometimes fitting their business processes to the software's best practices. Although some customization is possible during implementation, companies find they must still change some of the ways they work to fit the software.

By 1998, most of the Fortune 500 companies had already installed ERP systems, so ERP vendors refocused their marketing efforts on midsized companies (those with fewer than 1,000 employees). Midsized companies represented a ripe, profitable market. For example, midsized European companies have a total yearly budget of $50 billion for IT expenditures. The American market is even larger. SAP anticipates its business from midsized companies to rise to 40–45 percent of new orders by 2010, up from 30 percent in 2007. To appeal to this new market, SAP developed SAP All-in-One, a single package containing specific, preconfigured bundles of SAP ERP tailored for particular industries, such as automotive, banking, chemicals, and oil and gas. Because it is tailored to specific industries, SAP All-in-One can be installed more quickly than the standard ERP product.

At the time this book was published, SAP was working on the development of a new product, code-named "A1S." A1S is based on the SAP NetWeaver platform, and provides a solution designed specifically for fast-growing midsized companies with limited IT resources.

Application hosting, in which a third-party company provides the hardware and software support, rather than an internal IS department, is also making ERP systems like SAP more appealing to midsized companies.

SAP and Oracle also have some competition from smaller providers of ERP software. For example, Exact Software's ERP package, called e-Synergy, has similar functionality to the larger products, with seven modules: Human Resources, Document Managing, Financials, Logistics, CRM (Customer Relationship Management), Procurement, and Project. The company, headquartered in the Netherlands, targets small-to-midsized companies. Also, in 2000, software giant Microsoft acquired Great Plains, a provider of ERP software. The company's ERP products, now called Microsoft Dynamics, are designed for the small business market and offer a line of different software solutions. SAP has responded by creating SAP Business One, ERP systems for small business. It is interesting to note that while SAP and Microsoft are competing for the small business ERP market, they are also collaborating on other projects (see Chapter 8 for a discussion of the Duet product).

Responses of the Software to the Changing Market

In the mid-1990s, many companies complained about the difficulty of implementing the SAP R/3 system. SAP faced a canceled implementation by Dell Computers, a lengthy implementation at Owens Corning, and a lawsuit by the now-defunct FoxMeyer drug company. Cadbury experienced a surplus of chocolate bars at the start of 2006 in part due to a troubled ERP implementation. In response, SAP developed the Accelerated SAP (ASAP) implementation methodology, a framework for implementing systems, to ease the implementation process. SAP has continued developing implementation methodology; the latest version, Solution Manager, is designed to greatly speed the implementation process.

SAP continues to extend the capabilities of SAP ERP with additional, separate products that run on separate hardware and that extract data from the SAP ERP system. In many cases, these products provide more flexible and powerful versions of tools available in the SAP ERP system. The Business Warehouse (BW) product is an example of one such solution. Customers needing more capability and flexibility beyond the standard SD, PP, and FI modules can add BW. BW runs on a separate server and lets the user define unique reporting and analysis methods and integrate information from other systems. The downside of the BW product is that it represents another system for the user to purchase and implement. BW is now being referred to as BI, or Business Intelligence. At the moment, the two acronyms are interchangeable with reference to SAP. McKesson Pharmaceutical runs BI to find any inventory adjustments quickly and remedy the problem. At McKesson, products have high prices, low profit margins, and a limited shelf life. The company's BI system highlights problems immediately, not months after a product has expired.

The SAP ERP system provides some tools to manage customer interactions and analyze the success of promotional campaigns, but SAP sells a separate program called Customer Relationship Management (CRM), which has extended customer service capabilities. SAP's CRM product is designed to compete with CRM systems from competitors such as Siebel.

SAP addresses the issue of Internet-based data exchange with an integration platform called NetWeaver, which allows users to connect to SAP products through the Internet.

Like all technology, ERP software and related products are constantly changing. Thus, the challenge for a company is not only to evaluate an ERP vendor's current product offerings, but also to assess its development strategies and product plans.

CHOOSING CONSULTANTS AND VENDORS

Because ERP software packages are so large and complex, one person can't fully understand a single ERP system; it is also impossible for an individual to compare various systems. So, before choosing a software vendor, most companies study their needs and then hire an external team of software consultants to help choose the right software vendor(s) and the best approach to implementing ERP. Working as a team with the customer, the consultants apply their expertise to selecting an ERP vendor (or vendors) that will best meet their customer's needs.

After selecting a vendor, the consultants recommend the modules that are best suited to the company's operations and the configurations within those modules that are most appropriate. This preplanning should involve not only the consultants and a company's IT department, but the management of all functional business areas as well.

THE SIGNIFICANCE AND BENEFITS OF ERP SOFTWARE AND SYSTEMS

The significance of ERP lies in its many benefits. Recall that integrated information systems can lead to more efficient business processes that cost less than those in unintegrated systems. In addition, ERP systems offer the following benefits:

- ERP allows easier global integration: Barriers of currency exchange rates, language, and culture can be bridged automatically, so data can be integrated across international borders.
- ERP integrates people and data while eliminating the need to update and repair many separate computer systems. For example, Boeing had 450 data systems that fed data into its production process. The company now has a single way to record production data.

- ERP allows management to manage operations, not just monitor them. For example, without ERP, getting an answer to "How are we doing?" requires getting data from each business unit and then analyzing that data for a comprehensive, integrated picture. The ERP system already has all the data, allowing the manager to focus on improving processes. This focus enhances management of the company as a whole, and makes the organization more adaptable when change is required.

An ERP system can dramatically reduce costs and improve operational efficiency. For example, Rohm and Haas, the $8 billion chemical company, claims to have doubled revenue per employee over the past six years through productivity improvements due to an SAP implemention. These improvements can lead to lower costs and more satisified customers. Toyota anticipates savings of $7 million from its new ERP system, which allows employees to access their own human resource records and gives group leaders access to shop floor information on employees. Toyota is saving an additional $50,000 annually by providing access to HR software through a Web browser, meaning that the company does not have to install or support HR software on worker's desktops.

QUESTIONS ABOUT ERP

How Much Does an ERP System Cost?

Cost of an ERP system includes several factors:

- The size of the ERP software, which corresponds to the size of the company it serves
- The need for new hardware that is capable of running complex ERP software
- Consultants' and analysts' fees
- Time for implementation (which causes disruption of business)
- Training (which costs both time and money)

A large company, one with well over 1,000 employees, will likely spend $50 million to $500 million for an ERP system with operations involving multiple countries, currencies, languages, and tax laws. Such an installation might cost as much as $30 million in software license fees, $200 million in consulting fees, additional millions to purchase new hardware, and even more millions to train managers and employees—and full implementation of the new system might take four to six years.

A midsized company (one with fewer than 1,000 employees) might spend $10 million to $20 million in total implementation costs and have its ERP system up and running in about two years.

Should Every Business Buy an ERP Package?

ERP packages imply, by their design, a certain way of doing business, and they require users to follow that way of doing business. Some of a business's operations, and some segments of its operations, might not be a good match with the constraints inherent in ERP. Therefore, it is imperative for a business to analyze its own business strategy, organization, culture, and operation *before* choosing an ERP approach.

An article in the *Harvard Business Review* provides examples that show the value of planning: "Applied Materials gave up on its system when it found itself overwhelmed by the organization changes involved. Dow Chemical spent seven years and close to half a billion dollars implementing a mainframe-based enterprise system; now it has decided to start over again on a client-server version." In another example, Kmart in 2002 wrote off $130 million because of a failed ERP supply chain project. At the time, Kmart was not happy with its existing supply chain software, and it attempted to implement another product too quickly. In addition, the CIO left the company and new management wanted to make some changes.

The giant U.S. retailer Wal-Mart has chosen not to purchase an ERP system, but to write all its software in-house. Wal-Mart's philosophy is that the global strategic business process drives the technology. IT personnel are encouraged to consider the merchandising aspect of a process first and foremost, and then let the technology follow.

Sometimes, a company is not ready for ERP. In many cases, ERP implementation difficulties result when management does not fully understand its current business processes and cannot make implementation decisions in a timely manner. An advantage of an ERP system is that it can reduce costs by streamlining business processes. If a company is not prepared to change its business processes to make them more efficient, then it will find itself with a large bill for software and consulting fees, with no improvement in organizational performance.

Is ERP Software Inflexible?

Although many people claim that ERP systems, especially the SAP ERP system, are rigid, SAP ERP does offer numerous configuration options that help businesses customize the software to fit their unique needs. In addition, programmers can write specific routines for special applications in SAP's internal programming language, called **Advanced Business Application Programming (ABAP)**. The integration platform, NetWeaver, offers further flexibility in adding both SAP and non-SAP components to a company's IT infrastructure. Companies should be careful about how much custom programming they include in their implementations, because they might re-create their existing information systems in a new software package, instead of gaining the benefits of improved, integrated business processes. In its implementation of PeopleSoft, FedEx Corporation installed the systems for financial and HR functions with little or no modification.

Once an ERP system is in place, trying to reconfigure it while retaining data integrity is expensive and time-consuming. That is why thorough pre-implementation planning is so important. It is much easier to customize an ERP program during system configuration and before any data have been stored.

What Return Can a Company Expect from Its ERP Investment?

The financial benefits provided by an ERP system can be difficult to calculate because sometimes ERP increases revenue and decreases expenses in intangible ways that are difficult to measure. Also, some changes take place over such a long period of time that they are difficult to track. Finally, the old information system may not be able to provide good data on the performance of the company before the ERP implementation, making comparison difficult. Still, the return on an ERP investment can be measured and interpreted in many ways:

- Because ERP eliminates redundant effort and duplicated data, it can generate savings in operations expense. Because an ERP system can help produce goods and services more quickly, more sales can be generated every month.
- In some instances, a company that doesn't implement an ERP system might be forced out of business by competitors that have an ERP system—how do you calculate the monetary advantage of remaining in business?
- A smoothly running ERP system can save a company's personnel, suppliers, distributors, and customers much frustration—a benefit that is real, but difficult to quantify.
- Because both cost savings and increased revenues occur over many years, it is difficult to put an exact dollar figure to the amount accrued from the original ERP investment.
- Because ERP implementations take time, there may be other business factors affecting the company's costs and profitability, making it difficult to isolate the impact of the ERP system alone.
- ERP systems provide real-time data, allowing companies to improve external customer communications. Better communication can improve customer relationships and increase sales.

How Long Does It Take to See a Return on an ERP Investment?

A **return on investment (ROI)** is an assessment of an investment project's value, calculated by dividing the value of the project's benefits by the project's cost. An ERP system's ROI can be difficult to calculate because of the many intangible costs and benefits previously mentioned. Some companies do not even try to make the calculation, on the grounds that the package is as necessary as having electricity (which is not justified as an investment project). Companies that do make the ROI calculation have seen widely varying results. Some ERP consulting firms refuse to do ERP implementations unless their client company performs an ROI. Peerstone Research reported on over 200 companies using SAP or Oracle ERP systems and found that 38 percent of survey respondents do not do formal ROI evaluations.

In the Peerstone Research study, 63 percent of companies that did perform the calculation reported a positive ROI for ERP. Manufacturing firms are more likely to see a positive ROI than government or educational organizations. However, most companies felt that nonfinancial goals were the reason behind their ERP installations. Seventy-one percent of those companies surveyed said that the goal behind the ERP installation was improved management vision. Although Nestlé USA has had problems with its ERP implementation, it estimated a cost savings of $325 million, after spending six years and over $200 million on the implementation.

Toro, a wholesale lawnmower manufacturer, spent $25 million and four years to implement an ERP system. At first, ROI was difficult for Toro to quantify. Then, the emergence of an expanded customer base of national retailers, such as Sears and Home Depot, made it easier to quantify benefits. For example, Toro was able to gain a yearly savings of $10 million in inventory reduction—the result of better production, warehousing, and distribution methods.

Why Do Some Companies Have More Success with ERP Than Others?

Early ERP implementation reports indicated that only a low percentage of companies experienced a smooth rollout of their new ERP systems *and* immediately began receiving the benefits they anticipated. You should put such reports into perspective. *All* kinds of software implementations can suffer from delays, cost overruns, and performance problems, not just ERP projects. Such delays have been a major problem for the IS industry since the early days of business computing. Nevertheless, it is worth thinking specifically about why ERP installation problems can occur.

You can find numerous cases of implementation woes in the news. W. L. Gore, the maker of GoreTex, had some problems implementing its PeopleSoft system for personnel, payroll, and benefits. The manufacturer sued PeopleSoft, Deloitte & Touche LLP, and Deloitte Consulting for incompetence. W. L. Gore blamed the consultants for not understanding the system and leaving its Personnel department in a mess. PeopleSoft consultants were brought in to fix the problems, but the fix cost W. L. Gore additional hundreds of thousands of dollars.

Hershey Foods (now The Hershey Company) had a rough rollout of its ERP system in 1999, due to what experts say was the "Big Bang" approach to implementation, in which huge pieces of the system are implemented all at once. Companies rarely use this approach because it is so risky. Hershey lost a large share of the Halloween candy market that year due to ERP problems from this poor implementation. Hershey's order-processing and shipping departments had glitches that were being fixed as late as September that year.

Usually, a bumpy rollout and low ROI are caused by *people* problems and misguided expectations, not computer malfunctions. For example:

- Some executives blindly hope that new software will cure fundamental business problems that are not curable by any software. The root of the problem may lie in flawed core business processes. Unless the company changes its business processes, it will just be computerizing a bad way to do business.
- Some executives and IT managers don't take enough time for a proper analysis during the planning and implementation phase.
- Some executives and IT managers skimp on employee education and training.
- Some companies do not place the ownership or accountability for the implementation project on the personnel who will operate the system. This lack of ownership can lead to a situation in which the implementation becomes an IT project rather than a company-wide project.
- Unless a large project such as an ERP installation is promoted from the top down, it is doomed to fail. The top executives must be behind the project 100 percent for it to be successful.
- ERP implementation brings a tremendous amount of change for the users. Managers need to manage that change well so that the implementation goes smoothly.

Many ERP implementation experts stress the importance of proper education and training for both employees and managers. Most people will naturally resist changing the way they do their jobs. Many analysts have noted that active top management support is crucial for successful acceptance and implementation of such company-wide changes.

Some companies willingly part with funds for software and new hardware, but don't properly budget for employee training. ERP software is complex and can be intimidating at first. This fact alone supports the case for adequate training. Gartner Research recommends allocating 17 percent of the project's budget for training. Those companies spending less than 13 percent on training are three times more likely to have problems. The cost includes training employees on how to use the software to do their job, employees' nonproductive downtime during training, and—very important—educating employees about how the data they control affect the entire business operation.

Nestlé has learned many lessons from its implementation of ERP systems. Its six-year, $210 million project was initially headed for failure because Nestlé didn't include on the implementation team any employees from the operating groups affected. Employees left the company, morale was down, and help desk calls were up. After three years, the ERP implementation was temporarily stopped. Nestlé USA's vice president and CIO at that time, Jeri Dunn, learned that the project was not about implementing the software, but about change management. "When you move to SAP, you are changing the way people work. . . . You are challenging their principles, their beliefs and the way they have done things for many, many years," said Dunn. After addressing the initial problems, Nestlé ultimately reaped benefits from its ERP installation.

For many users, it takes years before they can take advantage of many of an ERP system's capabilities. Most ERP installations do generate returns, and news coverage now focuses on how companies gain value from their existing systems or are upgrading and adding functionality to their existing ERP systems. Del Monte Foods needed to meet Wal-Mart's and Target's requirements for package tracking using radio frequency identification devices (RFIDs), so approximately a year after its ERP system installation, the company tied its RFID applications into its existing SAP platform and is working to make the supply chain efficient.

ANOTHER LOOK

Digitizing the Depot: Beyond the No. 2 Pencil

When CIO Bob DeRodes joined Home Depot in 2001, he described the company's technology as consisting of the No. 2 pencil. Home Depot has come a long way since 2001. In 2005, the company was opening a new store every 24 hours; in the first six months of 2007, Home Depot opened 47 new stores. A No. 2 pencil couldn't keep up with all that expansion! A solid ERP system is the only type of system that could handle that growth. In fact, when asked if Home Depot could use open source software, DeRodes said, "Big companies need big software companies." In keeping with this statement, in 2002, Home Depot began a 10-year implementation of SAP software.

The implementation will help Home Depot in areas such as supply chain management and connecting electronically to its suppliers. SAP is already helping out in the four areas of Home Depot's expansion plans: international sales, Web sales, in-home installation, and sales to contractors. SAP's scalability is a good fit for Home Depot's expansion plans.

continued

Home Depot has some other lofty goals for the new system. The company will eventually allow each store employee to pull up detailed information about the product that is being sold. This information will include availability and uses. In addition, employees will have access to customer information. The product and customer data will be available at the checkout kiosk and also through a hand-held device.

Home Depot also uses business analytics to help manage pricing changes, which will differ from store to store. For example, if customers in one city are hesitant to buy a certain item, that particular item's price might be lowered for that store only. Also, Home Depot is embracing the service-oriented architecture of SAP's NetWeaver, as a way to link different software applications seamlessly.

Questions:

1. What type of information needs does a large and very rapidly expanding company like Home Depot have? Does a traditional ERP system meet those needs?

2. Research open source software. Why do you think the CIO implied that open source software is not appropriate for a large company like Home Depot?

THE CONTINUING EVOLUTION OF ERP

Understanding the social and business implications of new technologies is not easy. Howard H. Aiken, the pioneering computer engineer behind the first large-scale digital computer, the Harvard Mark I, predicted in 1947 that only six electronic digital computers would be needed to satisfy the computing needs of the entire United States! Hewlett-Packard passed up the opportunity to market the computer created by Steve Wozniak that became the Apple I. Microsoft founder Bill Gates did not appreciate the importance of the Internet until 1995, by which time Netscape controlled the bulk of the Internet browser market. (Gates, however, did dramatically reshape Microsoft around an Internet strategy by the late 1990s. Its Internet Explorer browser is more commonly used than any other.) Thus, even people who are most knowledgeable about a new technology do not always fully understand its capabilities or how it will change business and society.

ERP systems have been in common use only since the mid-1990s. As this young technology continues to mature, ERP vendors are working to solve the adaptability problems that plague customers. The demand for new ERP installations is still going strong. AMR Research has calculated that, in 2005, companies spent $14.5 billion on licenses and maintenance.

ANOTHER LOOK

Implementation Problems at Universities

ERP systems are attractive to universities for the same reasons that they are attractive to business organizations: control, accurate information, and centralized systems, all in real time. However, implementing ERP systems at universities has posed some unique problems. By nature, universities are not integrated organizations. Each department operates separately and autonomously, so trying to tie everyone together is difficult enough. Furthermore, university IT personnel are not as experienced as those in large companies, implementations are often rushed, and testing and training often have been inadequate.

Universities have been attracted to ERP systems since the mid-1990s, when their legacy systems were unable to keep up with increasing technology demands. There were too many legacy systems to maintain, and new systems couldn't be developed in-house because of a lack of staff and experience; ERP seemed like a good alternative. PeopleSoft aggressively marketed to universities and by the end of 2004 had 730 installations in colleges and universities.

Some universities have had a particularly difficult time with implementation. Stanford University began its PeopleSoft implementation in 2001. Stanford users complain that completing tasks takes longer than it did prior to the ERP system installation, while the university's IT department complains that the new system is more expensive to support than the prior system. Users at Stanford have been hesitant to adopt the new system, and many of those who are using it are requesting further customization. Lacking widespread use at this stage, the installation of this multimillion dollar system cannot be considered a complete success. Most of the problems related to the PeopleSoft installation are people problems; as with many corporate ERP implementations, the university and its IT department are coping with a tight budget and, consequently, providing little training. Although training was offered to the users, few participated.

The University of Massachusetts's PeopleSoft system was down for four days during the critical drop/add period in 2004, leaving 24,000 students in the lurch. The glitch was traced to a lack of testing. Cleveland State sued PeopleSoft because the college's software was "unusable" and the university had to install an alternative software package to process accounts receivables. In February 2005, Oracle, which now owns PeopleSoft, settled the suit for $4.25 million.

The University of Delaware has taken a slow approach to its PeopleSoft implementation. The first department to use the software was HR, for payroll. The university chose HR because the department is small. The next phase for the implementation was the university-wide financials. The last phase, and the largest phase, has been the student records. There have been a few hiccups, as in all implementations, but in general, it has gone smoothly.

Question:

1. Research the PeopleSoft corporation, and explain why its software is an attractive ERP package for higher education.

Additional Capabilities Within ERP

Sales production, data analysis, and Internet connectivity are a few areas where ERP vendors are expanding ERP capabilities. As discussed previously, ERP vendors and other software companies are continuing to develop Customer Relationship Management (CRM) applications that increase the efficiency of the sales force. For example, if salespeople can target the most profitable customers—actual and potential—they'll have a competitive advantage (you will learn more about CRM in Chapter 3). Other software focuses on detecting changes in customer satisfaction and responding quickly to remedy any problems.

Supply Chain Management (SCM) is one area that can benefit from ERP. SCM applications help to translate customer demand into production plans more efficiently. For example, SAP's Advanced Planner and Optimizer module provides sales representatives with a Global Available-to-Promise (ATP) capability. Global ATP allows the sales representatives to instantly check all available plants and warehouses in the company to find the best option for meeting a customer order in a timely and cost-efficient way. If an office supply company sends an order for 100 computer printers to a printer manufacturer, specifying delivery the following week, the manufacturer will look at its inventory of printers and production of printers. Traditionally, if the manufacturer has 100 printers in inventory, and 100 printers are to be manufactured the following week, the manufacturer will assign the items in inventory to the office supply company. This process handled in ATP would assign the future manufacturing run to the office supply company because the printer manufacturer could receive an additional order for 100 printers that needs to be filled immediately.

ERP developers are also trying to make their existing systems smarter by extending ERP's capabilities into more areas of decision support, management reporting, and data mining. **Data mining** is the statistical and logical analysis of large sets of transaction data, looking for patterns that can aid decision making. For example, discovering patterns in customer behavior can lead to better marketing efforts. Strategic Enterprise Management (SEM) applications help a company translate corporate-level goals (such as profit and market share targets) into operational decisions (such as production plans and workforce levels).

Internet connectivity is another area in which ERP capabilities are expanding. ERP vendors continue to improve software and Internet connections that integrate a business's internal operations, while also integrating the business with its dealers, vendors, and customers. Web services is one area that is receiving a lot of attention.

The Internet

The Internet's rapid development since the mid-1990s has been a threat to ERP software developers. ERP software lets users access the company's software and central database through internal connections. Now, users often need to access that central database directly from the Internet. Access through the Internet allows for greater flexibility in work because all one needs is a computer with an Internet connection and a Web browser. This accessibility has forced ERP companies to rethink how users get to and use their ERP software. ERP developers have been incorporating Web-based systems with their ERP products. This effort is typified by SAP's NetWeaver, an application that lets companies add components to their SAP ERP systems, and also allows external partners to access certain parts of the company's ERP system. Such open access is a necessity for the success of Internet-based business activities such as **electronic commerce** (or **e-commerce**), the conduct of business over the Internet. The Internet has become an important way to sell goods and services and will probably become more important in the future. Companies will have a continuing need to take orders electronically and to pass them seamlessly to the company's database. Once the order is in the database, the ERP system can manage the transaction as if it had come in through a traditional method, such as the telephone.

In conducting e-commerce, companies put a lot of focus on their sites' Web interfaces, spending great amounts of time and money to ensure that the site works smoothly. However, a company that intends to sell its products on the Web must still manage its business processes. Some experts have speculated that e-commerce will make ERP systems obsolete, but ERP is not likely to disappear. Rather, e-commerce is another activity that ERP systems can help manage.

As ERP installations continue to grow, so does interest in **Web services**, or, as it is frequently called, **service-oriented architecture (SOA)**. Web services are software that enables systems to exchange data without complicated software links. This capability allows companies to quickly launch new businesses, partnerships, and technology. A nonbusiness example of service-oriented architecture is the use of Craigslist and Google Maps to display rental property. These are two separate applications that work together on a third-party Web site. If you were searching Craigslist for an apartment in San Francisco, you could click the Google map link under the apartment listing and see exactly where it is located. Web services also make ERP systems easier to manage, especially when interfacing with other applications and the Web. For example, Apple was able to launch its iTunes store in three months by using SOA with SAP running as the back-office system. This shift from the traditional ERP client-server system to the service-oriented architecture is gaining momentum. SAP's NetWeaver platform serves the same role as SOA.

ANOTHER LOOK

Whirlpool and Web Services: Maximizing Value from an ERP System

Whirlpool is moving beyond the traditional ERP system into Web services for its Web site and phone support system. Using SAP's NetWeaver's service-oriented architecture, Whirlpool can be more flexible and responsive.

Whirlpool can outsource its Web page portal design because with service-oriented architecture, the Web page is separated from the underlying order-processing system. In fact, Whirlpool has been able to change its Web page in two to three days with this new flexibility, which extends to nonstandard devices such as PDAs.

In Whirlpool's phone support, three different areas are linked—the phone system, the product-tracking system, and the production system—which results in more efficiency. Note that these three areas originate from one in-house system and two different software systems.

Questions:

1. Research service-oriented architecture. Write a detailed definition and give an example.
2. Imagine you run a financial services company. How would service-oriented architecture help you in running your business?

Chapter Summary

Several factors led to the development of ERP:

- The speed and power of computing hardware increased exponentially, while cost and size decreased.

- The early client-server architecture provided the conceptual framework for multiple users sharing common data.

- Increasingly sophisticated software facilitated integration, especially in two areas: A/F and manufacturing resource planning.

- The growth of business size, complexity, and competition made business managers demand more efficient and competitive information systems.

- SAP AG produced a complex, modular ERP program called R/3. The software could integrate a company's entire business by using a common database that linked all operations, allowing real-time data sharing and streamlined operations.

- SAP R/3, now called SAP ERP, is modular software offering modules for Sales and Distribution, Materials Management, Production Planning, Quality Management, and other areas.

- ERP software is expensive to purchase and time-consuming to implement, and it requires significant employee training—but the payoffs can be spectacular. For some companies, however, the ROI may not be immediate or even calculable.

- Experts anticipate that ERP's future focus will be on managing customer relationships, improving planning and decision making, and linking operations to the Internet and other applications through service-oriented architecture.

Key Terms

Advanced Business Application Programming (ABAP)

Asset Management (AM) module

Best practices

Client-server architecture

Controlling (CO) module

Data mining

Database management system (DBMS)

Electronic commerce (e-commerce)

Electronic data interchange (EDI)

Financial Accounting (FI) module

Human Resources (HR) module

Legacy system

Materials Management (MM) module

Material requirements planning (MRP)

Modules

Open architecture

Plant Maintenance (PM) module

Production Planning (PP) module

Project System (PS) module

Quality Management (QM) module

R/3

Return on investment (ROI)

Sales and Distribution (SD) module

SAP ERP

Scalability

Service-oriented architecture (SOA)

Silo

Tolerance groups

Web services

Workflow (WF) module

Exercises

1. Define Moore's Law and explain why it is significant in the development of ERP. Is Moore's Law still holding?

2. What are the main characteristics of an ERP system? What are some newly developed features of ERP systems?

3. Imagine that you have been appointed the chief information officer of a start-up company that rents out DVDs via the Internet. An integrated information system is critical to your business. Write a proposal to the CEO highlighting the reasons why you need an ERP system. Use examples, gleaned from the Internet, of other companies' systems to augment your proposal.

4. Much has been written about ERP, both in the news media and on the Internet. Using library resources or the Internet, report on (1) one company's positive experience with implementing ERP, and (2) one company's disappointing experience.

5. Although ERP software packages have similar features, there are some differences between them. Document how Oracle's ERP systems and Microsoft's Dynamics differ from those of SAP. Visit www.oracle.com to research the "Applications" area, and visit www.microsoft.com.

6. Visit the online magazine CIO.com and conduct a search on ERP. Choose an example of an ERP implementation and write a memo to your instructor describing the procedure. Make comments as to areas in which the company could have improved its implementation.

For Further Study and Research

Allesch, Adolf. "Thriving or Surviving? How to Take Your SAP NetWeaver Pulse." *SAP NetWeaver Magazine* 03, no. 3 (Summer 2007). http://www.sap.com/community/pub/flash/kw28_07_story_2.epx.

Blau, John. "SAP arrives at Home Depot." *InfoWorld*, June 1, 2005. http://www.infoworld.com/article/05/06/01/HNsaphomedepot_1.html?SUPPLY%20CHAIN%20MANAGEMENT.

"Whirlpool whirls into Web services." *InfoWorld*, May 23, 2006. http://www.infoworld.com/article/06/05/23/78591_HNwhirlpoolwebservices_1.html?WEB%20SERVICES%20DEVELOPMENT.

Burleson, Donald. "Four factors that shape the cost of ERP." *TechRepublic*, August 16, 2001. http://search.techrepublic.com.com/index.php?q=Four+factors+that+shape+the+cost+of+ERP&t=11&go=Search.

"Connecting the Chemical Industry." *Chemical Week,* September 25, 2002.

Davenport, Thomas H. "Putting the Enterprise into the Enterprise System." *Harvard Business Review,* July–August 1998, 121–31.

Foroohar, Rana. "Software Savior?" *Newsweek*, January 29, 2007. http://www.msnbc.msn.com/id/16692270/site/newsweek/.

Few, Stephen. "The Information Cannot Speak for Itself." *IntelligentEnterprise.com,* July 10, 2004. http://www.intelligententerprise.com/showArticle.jhtml;jsessionid=0VWVEUEXBJOASQSNDLPCKH0CJUNN2JVN?articleID=22102226.

Forbes.com. "SAP sees orders from mid-sized businesses climbing to 45% from 30% by 2010." *Forbes.com*, April 13, 2007. http://www.forbes.com/afxnewslimited/feeds/afx/2007/04/13/afx3609185.html.

Greenbaum, Joshua. "The Ecosystem Advantage: It's Not Nice to Fool Mother Nature." *SAP NetWeaver Magazine* 03, no. 3 (Summer 2007). http://www.sapnetweavermagazine.com/archive/Volume_03_(2007)/Issue_03_(Summer)/v3i3a02.cfm?session=.

Hamerman, Paul and R. Wang. "ERP: Still a Challenge After All These Years." *Information Week,* July 11, 2005. http://www.informationweek.com/showArticle.jhtml?articleID=165600651.

Kirkpatrick, David. "The E-Ware War." *Fortune,* December 7, 1998, 102–112.

Kumar, Kuldeep and Jos van Hillegersberg. "ERP Experiences and Evolution." *Communications of the ACM*, 43, no. 4 (April 2000): 23–26.

MacDonald, Elizabeth. "W. L. Gore Alleges PeopleSoft, Deloitte Botched a Costly Software Installation." *The Wall Street Journal,* November 2, 1999.

McCue, Andy. "Too Much Candy: IT Glitch Costs Cadbury." *BusinessWeek*, June 8, 2006. http://www.businessweek.com/globalbiz/content/jun2006/gb20060608_252289.htm?chan=search.

Meissner, Gerd. *SAP: Inside the Secret Software Power.* New York: McGraw-Hill, 2000.

Montgomery, Nigel. "European Retailers Divided Over ERP Versus Best of Breed." *Tech Update,* April 18, 2003. http://techupdate.zdnet.com/techupdate/stories/main/0%2C14179%2C2913405%2C00.html.

Nadeau, Michael. "5 Keys to McKesson's Rapid BI Transformation." *SAP NetWeaver Magazine* 1, no. 1 (2005).

Nelson, Emily and Evan Ramstad. "Hershey's Biggest Dud Is Its New Computer System." *The Wall Street Journal,* November 4, 1999.

Nolan, Sean. "By the Numbers: November 2003." *Baseline.com,* November 1, 2003. http://www.baselinemag.com/article2/0,1540,1374440,00.asp.

Osterland, Andrew. "Blaming ERP." *CFO.com*, January 1, 2000. http://www.cfo.com/article.cfm/2987370.

Peerstone Research. "ERP ROI: Myth and Reality." *Information Week*, March 29, 2004. http://www.informationweek.com/reports/showReport.jhtml; jsessionid=5WZPDA3FLJXX0QSNDLQSKHSCJUNN2JVN?articleID=18600087&_requestid=16698.

PeopleSoft Case Studies. 2002. http://www.peoplesoft-hp.com/tools/successstories/%5Bfiles%5D/Enterprise/Automotive/Toyota%20Motor%20Manufacturing%20NA%20Drives%20Real-Time%20Productivity%20with%20PeopleSoft%208%20HRMS%20and%20Enterprise%20Portal.pdf.

SAP.com. "SAP Customer Success Story Chemicals: Rohm and Haas – MySAP Business Suite Aids in Drive for Six Sigma Business Process Improvement." SAP.com, 2007.

Schneider, Polly. "Human Touch Sorely Needed in ERP." *CIO,* March 2, 1999. http://www.cnn.com/TECH/computing/9903/02/erpeople.ent.idg/index.html.

Slater, Derek. "How to Choose the Right ERP Software Package," *CIO,* February 16, 1999. http://www.cnn.com/TECH/computing/9902/16/erppkg.ent.idg/index.html.

Sliwa, Carol. "IT difficulties help take Kmart Down." *Computerworld,* January 28, 2002.

Stein, Tom. "Making ERP Add Up." *Information Week,* May 24, 1999.

Steinert-Threlkeld, Tom. "Home Depot Hopes SAP Can Help Boost Sales." *Baseline,* May 18, 2005. http://www.baselinemag.com/article2/0,1397,1817341,00.asp.

Sullivan, Laurie. "ERPzilla." *InformationWeek*, July 11, 2005. http://www.informationweek.com/story/showArticle.jhtml?articleID=165700832.

van Everdingen, Yvonne, Jos van Hillegersberg, and Eric Waarts. "ERP Adoption by European Midsize Companies." *Communications of the ACM* 43, no. 4 (April 2000): 27–31.

Wailgum, Thomas. "University ERP: Big Mess on Campus." *CIO*, May 1, 2005. http://www.cio.com/article/107706/.

Westervelt, Robert. "U.S. software sales boost SAP earnings, says CEO." SearchSAP.com, April 22, 2004. http://searchsap.techtarget.com/originalContent/0,289142, sid21_gci961032,00.html.

Wheatley, Malcolm. "ERP Training Stinks." *CIO,* June 1, 2000.

White, Joseph B., Don Clark, and Silvia Ascarelli. "Program of Pain." *The Wall Street Journal,* March 14, 1997.

Whiting, Rick. "At SAP, R/3 Is a Distant Memory." *InformationWeek*, July 24, 2006. http://www.informationweek.com/story/showArticle.jhtml?articleID=190900725.

Whiting, Rick. "Home Depot Looks to SAP As It Modernizes." *InformationWeek*, May 23, 2005. http://www.informationweek.com/story/showArticle.jhtml?articleID=163106241.

Worthen, Ben. "Extreme ERP Makeover: How to Determine If a Single-Instance ERP Implementation Is Right for You." *CIO*, November 15, 2003. http://www.cio.com/article/31964/ How_to_Determine_If_a_Single_Instance_ERP_Implementation_is_Right_for_You.

Worthen, Ben. "Nestle's ERP Odyssey." *CIO*, May 15, 2002. http://www.cio.com/article/31066/ Nestle_s_Enterprise_Resource_Planning_ERP_Odyssey.

CHAPTER **3**

MARKETING INFORMATION SYSTEMS AND THE SALES ORDER PROCESS

LEARNING OBJECTIVES

After completing this chapter, you will be able to:

- Describe the unintegrated sales processes of the fictitious Fitter Snacker company.
- Explain why unintegrated Sales and Marketing information systems lead to company-wide inefficiency, higher costs, lost profits, and customer dissatisfaction.
- Discuss sales and distribution in the SAP ERP system, and explain how integrated data sharing increases company-wide efficiency.
- Describe how SAP ERP processes a standard sales order.
- Describe the benefits of customer relationship management software, a useful extension of ERP software.

INTRODUCTION

In this chapter, you will begin learning about the operations of Fitter Snacker (FS), a fictitious company that

makes healthy snack bars and does not have an integrated information system. Throughout the remainder

of this book, we will use FS to illustrate information systems concepts in general and ERP concepts in

particular.

As is the case in many other companies, Marketing and Sales (M/S) is the focal point of many of FS's activities. Why? Because Marketing and Sales is responsible for selling the company's product, meaning that marketing personnel often guide the company's key strategies and tactics. Marketing personnel in most companies make the following kinds of decisions:

- What products should we produce?

- How much of each product should we produce?

- How are our products best promoted and advertised?

- How should our products be distributed for maximum customer satisfaction?

- What price should we charge for our products?

On a day-to-day basis, M/S is involved in generating key transaction data (including data for recording sales), creating customers' bills, and allocating credit to customers. As you have learned in previous chapters, the availability of a common database to all modules is one of the advantages of having an integrated information system, because the data in the system are consistent. Integration can lead to problems, however; if the data are not correct, the error will carry over into all modules.

FS's M/S information systems are not well integrated with the company's other information systems. As a result, company-wide use of transaction data is inefficient, as you will see in this chapter. You'll also see how FS's M/S information systems could be improved by using ERP. We begin by looking at an overview of the company's operations.

OVERVIEW OF FITTER SNACKER

Fitter Snacker manufactures and sells two types of nutritious snack bars: NRG-A and NRG-B. NRG-A touts "advanced energy." NRG-B boasts "body building proteins." Each bar contains the following ingredients:

- Vitamins and Minerals: important nutrients

- Dry Base Mixture: oats, wheat germ, protein powder, and spices
- Wet Base Mixture: honey and canola oil

Each type of bar contains additional unique ingredients: NRG-A contains carob chips and raisins, while NRG-B contains hazelnuts and dates.

Fitter Snacker has organized its sales force into two groups, known as divisions: the Wholesale Division and the Direct Sales Division. The Wholesale Division sells to intermediaries who distribute the bars to small shops, vending machine operators, and health food stores. The Direct Sales Division sells directly to large grocery stores, sporting goods stores, and other large chain stores. Each division has an organizational structure that interacts with FS's other functional areas, such as Accounting and Supply Chain Management. The two divisions operate separately from one another, in effect breaking the M/S functional area into two pieces.

The two sales divisions differ primarily in terms of quantities of orders and pricing terms. The Direct Sales Division offers customers volume discounts to encourage larger sales orders, which are more efficient to process because each order—regardless of size—generates costs related to the paperwork, shipping, and handling of the order. Thus, an order of 500 cases of snack bars incurs the same handling costs as an order of 10 cases, but the large order might generate $5,000 in profit, while the small order might generate only $100. The Wholesale Division charges customers a lower fixed price because the orders are usually large (otherwise, the orders would be handled by the Direct Sales Division). Both divisions send their customers invoices requesting the total balance within in 30 days and offering a 2 percent discount if they pay within 10 days (2–10/net 30).

In addition to selling snack bars under the Fitter Snacker brand name, the company also packages the bars in store-brand wrappers for some chain stores.

PROBLEMS WITH FITTER SNACKER'S SALES PROCESS

Many of Fitter Snacker's sales orders have some sort of problem, such as incorrect pricing, excessive calls to the customer for information, delays in processing orders, missed delivery dates, and so on. These problems occur because FS has separate information systems throughout the company for three functional areas: the sales order system, the warehouse system, and the accounting system. Information from each system is shared electronically through periodic file transfers (sales order system to accounting system) and manually by paper printout (credit status from the Accounting department to sales clerks). The high number of transactions that are handled manually creates many opportunities for data entry errors. Further, all the information stored in the three systems is not available in real time, resulting in incorrect prices and credit information.

In each sales division, Fitter Snacker has four salespeople who work on the road, plus two sales clerks who work in the sales office. Salespeople work on commission and have some leeway in offering customers "discretionary discounts" to make a sale. The entire sales process involves a series of steps that require coordination between Sales, Warehouse, Accounting, and Receiving, as shown in Figure 3-1 on the next page. (Notice that Manufacturing is usually not directly involved in the sales process because goods are shipped from the warehouse.)

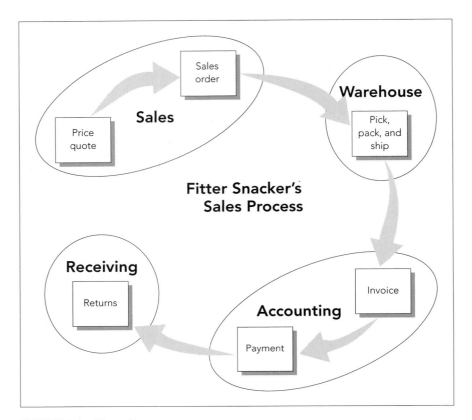

FIGURE 3-1 The sales process

Sales Quotations and Orders

Giving a customer a price quotation and then taking the customer's order should be a straightforward process, but at FS it is not. For a new customer, the sales process begins with a sales call, during which the salesperson either telephones the customer or visits in person. At the end of the sales call, the salesperson prepares a handwritten quotation on a form that generates two copies. The original sheet goes to the customer, the middle copy is first faxed and then mailed to the sales office, and the salesperson keeps the bottom copy for his or her records. The quotation form has an 800 number that the customer can call to place an order.

A number of problems can occur with this process:

- The salesperson might make an arithmetic error in the sales quotation. For example, a salesperson in the Direct Sales Division might offer both a quantity discount and a discretionary discount. If the salesperson isn't careful, the two discounts combined might be so deep that the company receives little or no profit.
- Salespeople fax a copy of their sales quotations to the sales office, but sometimes the same customer calls to place an order before the fax is transmitted.

The in-office clerk has no knowledge of the terms of the sale (which are outlined on the quotation) and must ask the customer to repeat the information. On the other hand, even if the quotation has been faxed, the data might not have been entered into the customer database, and still the customer must repeat the order information, much to his annoyance. This situation can also lead to a duplicate order.

- The fax received by the sales office is a faxed copy of a handwritten form, and might not be legible.

When customers place an order, they usually ask for a delivery date. To get a shipping date, the in-office clerk must contact the warehouse supervisor and ask whether the customer's order can be shipped from inventory, or whether shipping will be delayed until a future production run is delivered to the warehouse. Because she's too busy to make an inventory count, total all orders waiting to be filled, and find out how many orders are in process in the sales office, the warehouse supervisor can only estimate the shipping date.

Once the in-office sales clerk has the warehouse supervisor's estimated shipping date, she determines the shipping method for the order and how long delivery will take. Next, the clerk checks the customer's credit status. For new customers, the clerk fills out a paper credit-check form that includes basic customer data and the amount of the order. The form goes to Accounting, where accountants perform the credit check and then return the credit-check form showing the customer's credit limit. If the credit limit is below the amount of the purchase, assuming there are no other orders outstanding, the clerk calls the customer to determine what action the customer wants to take (reduce the amount of the order, prepay, or dispute the amount of credit granted). If the order is from an existing customer, the clerk checks a paper report from Accounting that shows the customer's current balance, credit limit, and available balance. Because the report is generated weekly, it might not reflect a customer's most recent payments.

The sales clerk enters the customer's order into the current order entry system. The computer program performs four important tasks. First, it stores the customer's order data, which are used later to analyze sales performance at the division level. Second, it prints out a packing list and shipping labels for the warehouse to use to pick, pack, and ship the customer's order. Third, it produces a data file of all current transactions for the Accounting department to use for preparing invoices (this file is also used for financial, tax, and managerial accounting, which is discussed in Chapter 5). And fourth, the data file is copied to a disk and entered into a PC-based accounting software program on Mondays, Wednesdays, and Fridays.

Order Filling

Fitter Snacker's process for filling an order is no more efficient than its sales order process. Packing lists and shipping labels are printed twice a day—at noon and at the end of the day. These are hand-carried to the warehouse, where they are hand-sorted into small orders and large orders. The Production department produces and wraps the bars and packs them in display boxes, 24 bars to a box. The display boxes have promotional printing and are designed to serve as a display case. FS packs 12 display boxes together to form a standard shipping case. Depending on the inventory levels in the warehouse, Production personnel might transfer the display boxes directly to the warehouse, or they might pack the display boxes into shipping cases.

The warehouse stores both display boxes and shipping cases, organized by label type (FS brand and store brand). For small orders (less than a full shipping case), the order picker goes to the warehouse with a handcart and pulls the number of display boxes listed on the packing list. If there are not enough display boxes in the warehouse to fill the order, the picker might break open a shipping case to get the required number of display boxes. If he does this, he is supposed to advise the warehouse supervisor so she can update the inventory records—but sometimes this step is overlooked.

The picker then brings the display boxes back to the small-order packing area, where they are packed into a labeled box with the packing list enclosed and prepared for shipping by a small package shipper.

For large orders (one or more shipping cases), the picker uses a forklift to move the appropriate number of shipping cases to the large-order packing area. Workers label them for shipping, load them on a pallet, and attach them to the pallet with shrink-wrap plastic for protection. These pallets are shipped either by one of FS's two delivery trucks or by a less-than-truckload (LTL) common carrier.

FS uses a PC database program to manage inventory levels in the warehouse. The program adjusts inventory level figures on a daily basis, using production records (showing what has been added to the warehouse), packing lists (showing what has been shipped from the warehouse), and any additional sources of data (such as shipping cases that have been opened to pull display boxes). Each month the warehouse staff conducts a physical inventory count to compare the actual inventory on hand with what the inventory records in the PC database show. Fitter Snacker's monthly inventory counts show that inventory records are more than 95 percent accurate. Ninety-five percent accuracy doesn't sound too bad, but FS still has problems filling orders. These are described next.

Because snack bars are somewhat perishable, FS keeps inventory levels fairly low, and inventory levels change rapidly during the day. As a result, a picker might go to the shelves to pick an order and discover that there are not enough of the desired type of snack bars to fill the order. The picker then has to decide from among several courses of action. There might be more of that type of bar in the production area, on the way to the warehouse. For an important customer, the wrappers and display box labels on the production line might be changed to the customer's brand to produce enough bars to complete the order. In other situations, the customer may want a partial shipment consisting of whatever is on hand, with the rest shipped when it becomes available, which is known as a backorder. Or, the customer might prefer to take the goods on hand, cancel the balance of the order, and place a new order later. If the customer's company has enough inventory on hand, the customer may wait until the whole order can be shipped, thus saving on delivery charges.

To determine what to do in this situation, the order picker might have conversations with the warehouse supervisor, production supervisor, and sales clerks. Whatever the final decision, the warehouse supervisor has to contact the sales clerk so she can notify the customer (which doesn't always happen, when things are busy) and the Accounting department so they can change the invoice.

Accounting and Invoicing

Invoicing the customer is problematic as well. Three times a week, sales clerks send the Accounting department the disk containing the sales order data for customer invoices. The Accounting department loads the data into the PC-based accounting program; then, clerks manually make adjustments for partial shipments and any other changes that have occurred during the order process. Sometimes, order corrections are delayed and don't catch up to the invoicing process, resulting in late or inaccurate invoices. If the completed invoice is waiting to be mailed when the warehouse notifies Accounting of a partial shipment, then a new invoice must be prepared. In any case, an invoice is eventually sent to the customer, separate from the shipment.

Payment and Returns

Fitter Snacker's procedure for processing payments can yield frustrating results for customers. Almost all customers pay the invoice within 10 days to receive the 2 percent discount. If any errors have occurred in the sales process—from the original quotation to entering the order into the sales order program—the customer will receive an incorrect invoice. Even though FS provides customers with two invoice copies, many customers don't return a copy of the invoice with their payment, as instructed. Errors result if the correct customer's account isn't credited.

FS's returns processing is also flawed. Because FS's snack bars contain no preservatives, they have a relatively short shelf life. Thus, FS has a policy of crediting customer accounts for returned snack bars that have exceeded their "sell by" date (this is a generous policy, because it is impossible to know who—FS or the customer—is responsible for the bars not selling before they expire). FS also gives credit for damaged or defective cases. Customers are supposed to call FS to get a returned material authorization (RMA) number to simplify the crediting process. When cases are returned to FS, the Receiving department completes a handwritten returned material sheet, listing the returning company's name, the materials returned, and the RMA number. However, many customers do not call for the RMA number, or fail to include it with their returned material, which makes it more difficult for the Accounting department to credit the appropriate account. Poor penmanship on the returned material sheet also creates problems for Accounting.

When an account becomes past due, FS sends a dunning letter, notifying the customer that the account is past due and requesting payment if payment hasn't already been sent. As the account gets more delinquent, the dunning letters usually get more direct and threatening. If a customer's account has not been properly credited, however, the customer may receive a dunning letter in error, or may receive a call about exceeding their credit limit after placing a new order. Such situations damage goodwill with both new and repeat customers.

In the following sections, you will learn how an ERP system can improve the sales process for Fitter Snacker.

SALES AND DISTRIBUTION IN ERP

An ERP system can improve the sales order process in several ways. Because ERP systems use a common database, they can minimize data entry errors and provide accurate information in real time to all users. An ERP system can also track all transactions (such as invoices, packing lists, RMA numbers, and payments) involved in the sales order.

Let's look at how one ERP system, SAP's ERP and its Sales and Distribution module, manages the sales order process. Other ERP software handles the process in a similar fashion. In SAP ERP, important transactions and events are assigned a number for record-keeping purposes. The electronic evidence of a transaction in SAP ERP is called a "document."

The SAP ERP Sales and Distribution module treats the sales order process as a cycle of events. SAP ERP defines up to six events for any sale:

- Pre-sales activities
- Sales order processing
- Inventory sourcing
- Delivery
- Billing
- Payment

Pre-Sales Activities

The first step in the SAP ERP sales and distribution process is Pre-Sales Activities. At this phase, customers can get pricing information about the company's products, either through an inquiry or a price quotation. The difference between an inquiry and a quotation is that a quotation is a written, binding document; the seller guarantees the buyer that, for some specified period of time, he can buy the product at the quoted price. The inquiry is simply a statement of prices, with no guarantee implied.

Pre-sales activities also include marketing activities such as tracking customer contacts, including sales calls, visits, and mailings. The company can maintain data about customers and generate mailing lists based on specific customer characteristics, which enhances targeted marketing efforts.

Sales Order Processing

In the SAP ERP system, sales order processing is the series of activities that must take place to record a sales order. The sales order can start from a quotation or inquiry generated in the pre-sales step. Any information that was collected from the customer to support the quotation (contact name, address, phone number) is immediately included in the sales order.

Some of the more critical steps in sales order processing are recording the items to be purchased, determining the selling price, and recording the order quantities. Users can define various pricing alternatives in the SAP ERP system. For example, a company can use product-specific pricing, such as quantity discounts, or it can define discounts that depend on both the product and a particular customer. Configuring a complex pricing scheme requires a significant amount of programming work, but once the system is in place, it will automatically calculate the correct price for each customer, eliminating many problems that FS experiences.

During sales order processing, the SAP ERP system checks the Accounts Receivable tables in the SAP ERP database to confirm the customer's available credit. SAP ERP adds the value of the order to the credit balance, and then compares the result to the customer's credit limit (also available in the database). If the customer has sufficient credit available, the order is completed. If not, the SAP ERP system prompts sales personnel to either reject the order, call the customer to check on recent payments, or contact Accounting to discuss any extenuating circumstances.

Inventory Sourcing

When recording an order, the SAP ERP system checks the company's inventory records and the production planning records to see whether the requested material is available and can be delivered on the date the customer desires. This Available-to-Promise (ATP) check includes the expected shipping time, taking into account weekends and holidays. Fitter Snacker's current system does not provide a good method for checking inventory availability. In the SAP ERP system, availability is automatically checked and the system can recommend an increase in planned production if a shortfall is expected. SAP also keeps a record of all open orders, so even if the ordered product is still in the warehouse, the system will reserve it on the shelf, making it unavailable to other customers.

Delivery

In the SAP ERP system, the word **delivery** means releasing the documents that the warehouse uses to pick, pack, and ship orders, rather than the traditional definition of transferring goods. The delivery process allows deliveries to be created so that the warehouse and shipping activities are carried out efficiently (for example, combining similar orders for picking, or grouping orders based on shipping method and destination).

Once the system has created the documents for picking, packing, and shipping, the documents are transferred to the Materials Management module, where the warehouse activities of picking, packing, and shipping are carried out.

Billing

Next, the SAP ERP system creates an invoice by copying the sales order data into the invoice document. Accounting can print this document and mail it, fax it, or transmit it electronically to the customer. Accounting records are also updated at this point. To record the sale, SAP ERP debits (increases) Accounts Receivable and credits Sales, thus updating the accounting records automatically.

Payment

When the customer sends in a payment (physically or electronically), it is automatically processed by the SAP ERP system, which debits cash and credits (reduces) the customer's account. Notice that the timely recording of this transaction has an effect on the timeliness and accuracy of any subsequent credit checks for the customer. Fitter Snacker has had a problem with getting accurate credit checks, frequently blocking orders for companies that are within their credit limit, while granting credit to other companies beyond what is advisable.

A STANDARD ORDER IN SAP ERP

Now we will take a look at how Fitter Snacker's sales order process would work with an SAP ERP system in place. You will learn how this ERP system would make FS's sales order process more accurate and efficient. Notice that ERP allows business processes to cut across functional area lines.

Taking an Order in SAP ERP

Figure 3-2 shows an order entry screen in SAP ERP's 4.7 Enterprise system, the version of the software released in 2003. The important fields in this screen are summarized in Figure 3-3.

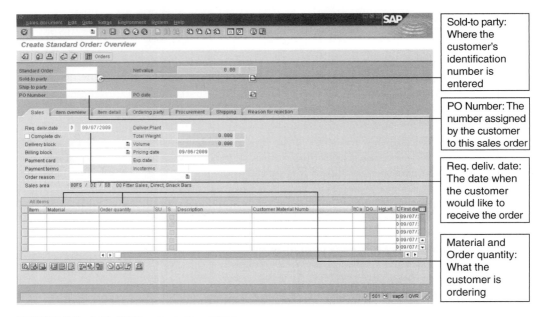

FIGURE 3-2   SAP ERP order entry screen

In SAP ERP, a sales order clerk must enter code numbers for customers' names and the inventory sold, rather than using customers' names and inventory item names. Because more than one customer might have the same name, a unique number is assigned by the company to each customer in the database. This number acts as the primary identifier for the customer. The same logic applies to distinguishing one inventory item from another. In database terminology, such codes are called key fields. While it sounds as if the sales order clerk must remember a lot of code numbers to use the SAP ERP system, this is not the case. For most data entry fields, the SAP ERP system determines whether an entry is valid. Figure 3-4 shows how SAP ERP lists the system's 34 predefined sales order types. The sales order clerk can choose the correct type from among these.

Data Entry Field	Explanation
Sold-to party	Identification number assigned to customer
P.O. Number	The number assigned by the customer to the sales transaction. This is different from the sales order number assigned by the Seller (using SAP ERP) to the sales transaction. In a paper process, the purchase order number is usually a sequential number pre-printed on the purchase order form.
Req. deliv. date	The delivery date for the order requested by the customer. The SAP ERP system will evaluate the ability to meet this date and suggest alternatives, if necessary.
Material	The identification number assigned in the SAP ERP system to the item requested by the customer.
Order quantity	The number of units of the material the customer is requesting.

FIGURE 3-3   Data entry fields in the order entry screen

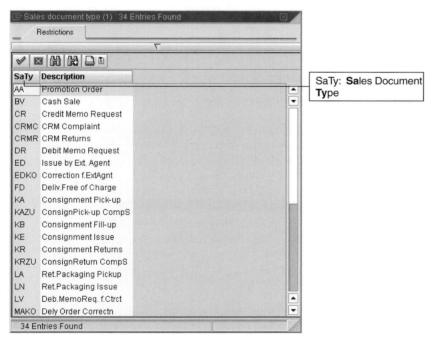

FIGURE 3-4   Some of the sales order (document) types predefined in SAP ERP

When a sales order clerk has to find a customer in the SAP ERP system, he or she can click the Sold-to party field to display a search icon and then click the search icon to open the sophisticated search window shown in Figure 3-5. With this search window, the clerk can search for a customer using different criteria (for example, part of the customer's address, such as City) to narrow the list produced by the search. Conducting a search using the criteria of Distribution Channel (specifying direct distribution, shown as DI, in the figure) and Division (specifying snack bar sales division, shown as SB, in the figure) produces a list of customers (Figure 3-6) from which the clerk can pick the correct customer.

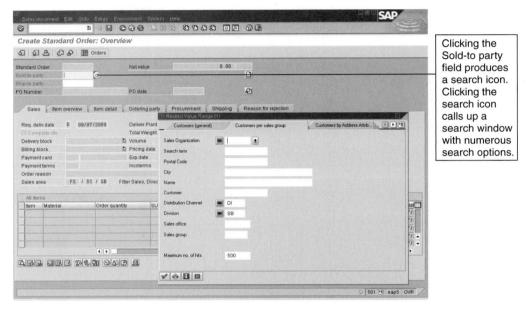

FIGURE 3-5 Search screen for customers

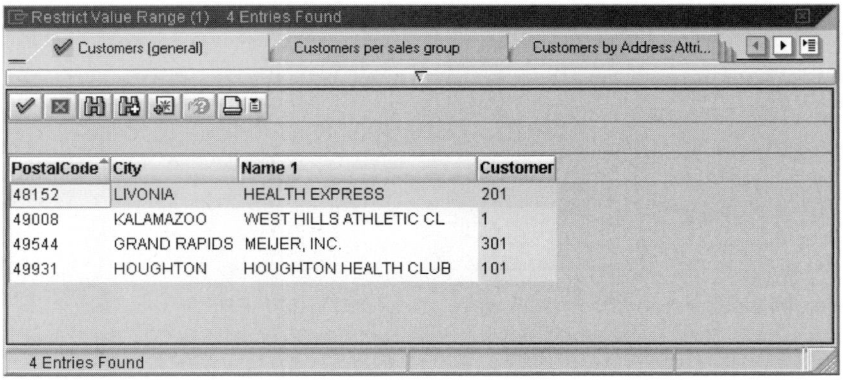

FIGURE 3-6 Results of customer search

In the SAP ERP system, a range of information is stored about each customer in multiple tables. These data are referred to as **customer master data**. Master data are data that remain fairly stable, such as customer name and address. Master data are maintained in the central database, and are available to all SAP ERP modules, including SD (Sales and Distribution), FI (Financial), and CO (Controlling).

Information about materials is in tables collectively called **material master data**, which are used by the MM (Materials Management—purchasing and warehousing) and PP (Production Planning) modules in addition to the SD module.

The SAP ERP system allows the user to define various ways to group customers and salespeople. These groupings are called **organizational structures**. One important organizational structure for Fitter Snacker is the Distribution Channel. With SAP ERP, the Distribution Channel allows the user to define different ways for materials to be sold and distributed to the customer. It allows for different types of relationships with different customers, and lets the company specify aspects of the relationship, such as pricing, delivery method, and minimum order quantities. Defining a Wholesale Distribution Channel and a Direct Sales Distribution Channel for FS would help to ensure that customers' orders are correctly priced.

Figure 3-7 shows a completed sales order screen for an order from West Hills Athletic Club for 10 cases of NRG-A bars and 10 cases of NRG-B bars.

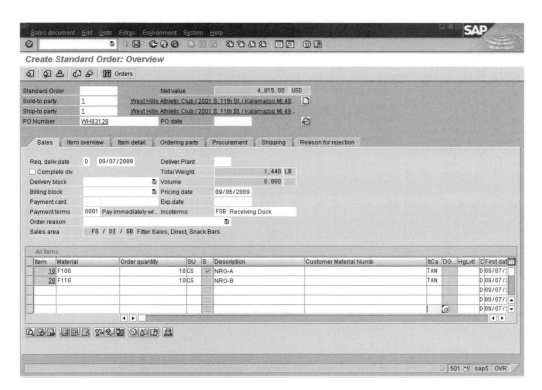

FIGURE 3-7   Order screen with complete data

West Hills's purchase order (PO) number for this sales transaction is WH83128. This is a unique number provided by the customer (West Hills) that allows it to track orders with its suppliers. If the customer has a question about an order, it will reference that PO number in its inquiry. Because the SAP ERP system records the customer's PO number, FS can look up the status of the order using the customer's PO number in addition to the sales order number that SAP will assign to this transaction.

Notice also that the order screen shows the name of the customer (West Hills Athletic Club) and the names of the products (NRG-A and NRG-B), even though they were not entered directly during the order process. How did they get on the screen at this point? Once the company code is entered, the clerk can easily request the SAP ERP system to search the database and access all company information needed to complete the order. Thus, the SAP ERP system simplifies the data entry tasks, reducing data entry time and the possibility of error.

When the SAP ERP system is instructed to save a sales order, it performs inventory sourcing—that is, it carries out checks to ensure that the customer's sales order can be delivered on the requested delivery date. Remember that previously when an FS customer wanted to know when an order could be delivered, the sales clerk had to make a series of phone calls. In the SAP ERP system, this checking is done automatically. It includes checks on both inventory and production, and it includes the time required for shipping. When the requested delivery date cannot be met, the SAP ERP system automatically proposes alternatives to the sales order clerk. For example, Figure 3-8 shows what the sales order clerk will see if only five cases of NRG-A bars will be available by the requested delivery date. The SAP ERP system provides the sales order clerk with three alternatives:

1. Reduce the order quantity of NRG-A cases to five, which can be delivered by the requested delivery date.
2. Delay shipment of the NRG-A bars for three days, when 10 cases will be available.
3. Ship five cases by the requested delivery data, and ship five cases at the later date.

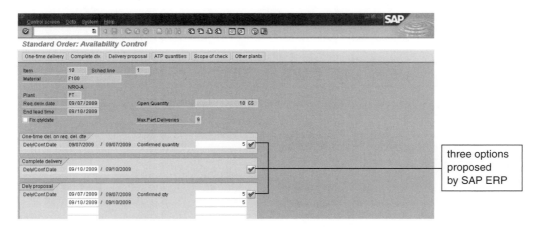

FIGURE 3-8 Order proposals

The best situation would be to meet the customer's request, but when that is not possible, the integrated SAP ERP system can provide the clerk with accurate information about the ability to meet the customer's requirements. With Fitter Snacker's previous unintegrated systems, it could take days for the customer to find out that its requested delivery date could not be met.

When a sales order is saved, the SAP ERP system assigns a document number to the sales order transaction. In the SAP ERP system, a document is an electronic record of a business transaction, and a document number is created for each business transaction. When the sales order is ready to be processed by the warehouse, a delivery document will be created with its own unique document number, which the system will link to the sales order document. Finally, when the bill (invoice) is prepared for the customer, the bill's unique number (called the invoice number) will be created and related to all the other numbers associated with the sales order.

The SAP ERP system has a mechanism for keeping track of the document numbers for the sales order so that employees can track the status of an order while it is in process, or research it after shipping. The linked set of document numbers related to an order, the **audit trail**, is called a **document flow** in SAP ERP. Figure 3-9 shows the document flow for a completed order. If an order includes partial shipments, partial payments, and returned material credits, the document flow can become quite complex. Without an integrated information system, the audit trail can be hard to establish, especially if many paper documents are involved. With an integrated system such as SAP ERP, document numbers are all linked together electronically. Not only does the document flow show all documents related to a sales order, but the user can look at the details of each document simply by double-clicking a line in the document flow. For example, if West Hills Athletic Club chose to take delivery of five cases of NRG-A bars on its requested delivery date and five cases when they became available, a sales order clerk can easily check on the status of the five delayed cases using document flow. He or she can search the SAP system to find the original sales order by a number of methods (open orders for West Hills, the PO number that West Hills used, the sales order number assigned by the SAP ERP system, and so on), and then review the delivery document for the delayed cases. As another example, a customer may have a question about an invoice it has received. The clerk can search the SAP ERP system for the invoice; the document flow will show all activity that led to the invoice. With an unintegrated information system, researching the invoice would require checking more than one information system and perhaps searching paper records as well.

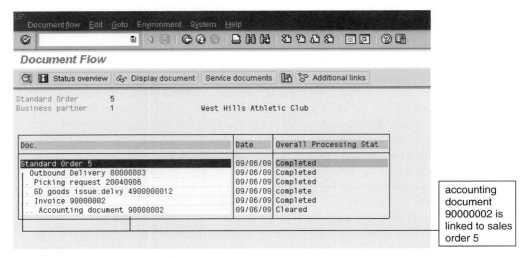

FIGURE 3-9   The Document Flow tool, which links sales order documents

Discount Pricing in SAP ERP

When a company installs an ERP system, it can configure it for a number of pricing strategies. For example, various kinds of discounts can be allowed (per item, on all items, based on unit price, based on total value, with or without shipping charges and taxes, by individual customer, by a class of customer, and so on). As a safeguard, the system can enforce limits on the size of discounts, to keep salespeople from offering unprofitable or unapproved discounts.

Suppose a salesperson wants to give a certain customer a 10 percent discretionary discount on NRG-A bars. But is the salesperson allowed to discount those bars? Is that discount appropriate for that customer? If so, will the discount be so deep that the sale will be unprofitable for FS? An ERP system automatically answers these questions.

Pricing, the process of determining how much to charge a particular customer, can be very complex. To accommodate the various ways that companies offer price discounts, SAP has developed a control mechanism it calls the **condition technique**. While detailed discussion of the condition technique is beyond the scope of this text, Figure 3-10 shows how a discount is automatically applied in the SAP ERP system.

The screen shows the price that FS is offering to West Hills Athletic Club for 10 cases of NRG-A bars. The base price is $240/case, and West Hills is being given a 10 percent discount automatically.

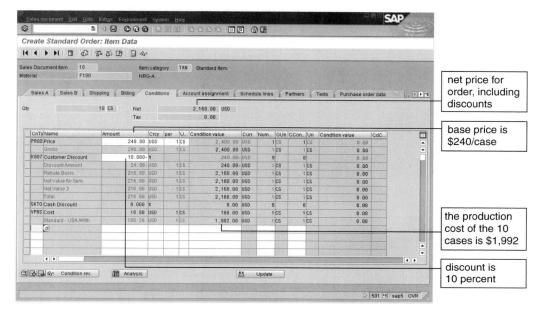

FIGURE 3-10 Pricing conditions for sales order

The screen in Figure 3-11 shows that the SAP ERP system has been configured to give West Hills Athletic Club a 5 percent discount for ordering over $1,000 worth of a particular type of snack bar, and a 10 percent discount if the order for a particular type of snack bar is over $1,500 in value.

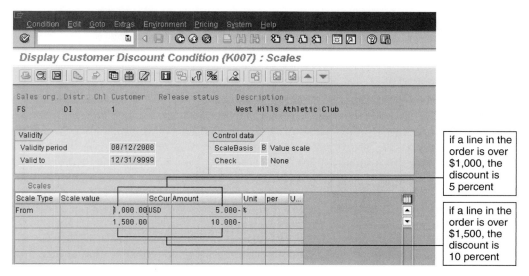

FIGURE 3-11 West Hills Athletic Club price discount

If you look again at Figure 3-10, you can see that the SAP ERP system automatically applied the 10 percent discount, so that the net price for 10 cases of NRG-A bars is $2,160. Figure 3-10 also shows that Fitter Snacker will still make a profit on this order, because the cost to produce 10 cases of NRG-A bars is $1,992.

Integration of Sales and Accounting

In the previous FS system, sales records were not integrated with the company's accounting records, so Accounting information was not always up-to-date. By contrast, ERP systems integrate Accounting with all business processes, so that when a sales order is recorded, the related accounting data are updated automatically.

In the document flow shown in Figure 3-9, Accounting document 90000002 is part of the sales order process. The clerk can select this line, then click the Display document button and get the accounting detail shown in Figure 3-12.

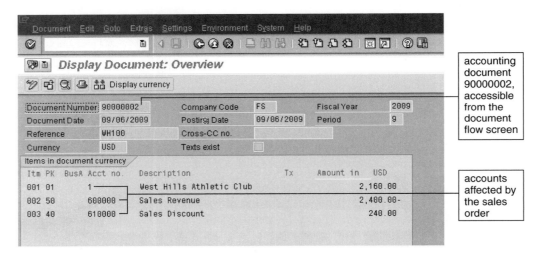

accounting document 90000002, accessible from the document flow screen

accounts affected by the sales order

FIGURE 3-12 Accounting detail for the West Hills sales order

This document shows that, as a result of this sales order, West Hills Athletic Club has paid Fitter Snacker $2,160. The Sales Revenue account (account number 600000) records sales revenue of $2,400 dollars, which represents the list price of the snack bars West Hills Athletic Club purchased. The 10 percent discount of $240 is recorded in the Sales Discount account (account number 610000). Because the accounting documents are created automatically with the sales order, the Accounting department is using the same data as Sales, which results in up-to-date and accurate information. If Marketing needs a report on the size of discounts currently being offered, the system can easily generate a report of that data, and it will be up-to-date and accurate as well.

Now that we have looked at how an ERP system can help a company's sales order process, let's look at how data obtained during that process might be further used to help Fitter Snacker build long-term business relationships with its customers and improve business performance.

CUSTOMER RELATIONSHIP MANAGEMENT

Companies without a good connection between their workers and their customers run the risk of losing business. Take Fitter Snacker, for example. Let's say a salesperson calls on a good customer, Health Express. The salesperson offers Health Express deep discounts for buying a certain number of NRG bars. At the same time, the Marketing department is running a sale on NRG bars. Marketing sends Health Express a flyer advertising the discounted price, which is less than what the salesperson just offered in person. Meanwhile, the vice president of Fitter Snacker plays golf with the CEO of Health Express and offers yet another discount. The connection or relationship with the customer is confused. **customer relationship management (CRM) software** can help companies streamline their interactions with customers.

Companies with an ERP system have an added benefit beyond systems integration: vast and complete quantities of data available for analysis. By adding other software tools to its ERP system, a company can extend the capabilities of the system, thus increasing its value. Many ERP vendors—and non-ERP software companies—provide CRM software. Siebel (bought by Oracle in 2005 for $5.8 billion) led the CRM software market until 2004, when its market dominance was overtaken by SAP. According to SAP, 34,000 companies are using SAP CRM. Forrester Research estimates the market for CRM software will reach $10.9 billion by 2010.

The latest method of delivering this software is on-demand. According to AMR Research, 12 percent of the market (amounting to $600 million) uses CRM by this method. With **on-demand** CRM, the software and computer equipment reside with the CRM provider; it is not installed in-house. Salesforce.com, for example, is a company that provides CRM software as it is needed by the user. On-demand access is provided by SAP and other vendors, including Oracle. Microsoft is also in the mix with its Dynamics CRM 3.0. Vendor pricing varies from $65 to $125 per user per month.

With the goal of providing "a single face to the customer," the basic principle behind CRM is that any employee in contact with the customer should have access to all information about past interactions with the customer. In a traditional sales organization, information might only be available to one individual, who might not share it with the organization. If that individual leaves the company, the information is lost.

Core CRM Activities

In general, all CRM software supports the following activities and tools:

- One-to-one marketing: Once a customer is categorized, the company can tailor products, promotions, and pricing accordingly. Customers can be offered products related to what they are now buying (cross-selling) or higher-margin products in the same lines (up-selling).
- Sales force automation (SFA): Occurrences of customer contacts are logged in the company's database. SFA software can automatically route customers who contact the company to a sales representative. Companies can use SFA software to forecast customer needs, based on the customer's history and transactions, and to alert sales representatives accordingly. Sometimes this software is called "lead management software" because a transaction can be tracked from the initial lead to post-sale follow-up.
- Sales campaign management: This software lets a company organize a marketing campaign and compile its results automatically.
- Marketing encyclopedias: This software serves as a database of promotional literature about products. The material can be routed to sales representatives or customers as needed.
- Call center automation: When customers call a company to get assistance with a company's products, representatives can query a knowledge management database containing information about the product. Some knowledge management software accepts queries in natural language. If the company must develop a new solution in response to a unique customer query, that information can be added to the knowledge base, which thus becomes "smarter."

SAP's CRM Software

A number of tools that provide CRM functionality exist within the SAP ERP system. For example, to make sure that information about sales contacts is available throughout the organization, the SAP ERP system provides a contact management tool, shown in Figure 3-13. This tool is essentially a database of personal contact information.

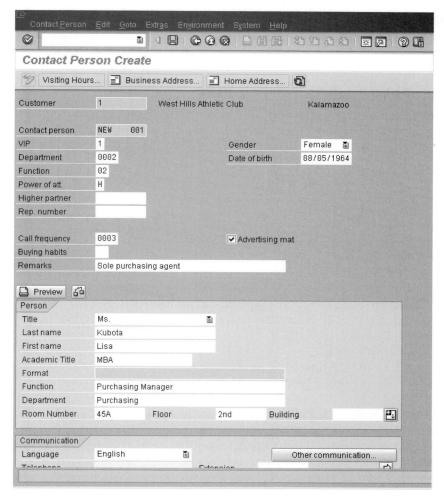

FIGURE 3-13   SAP ERP contact manager

Another CRM tool in the SAP ERP system is sales activity manager, an example of which is shown in Figure 3-14. This tool supports a strategic and organized approach to sales activity planning, and most importantly, can help make sure that follow-up activities are accomplished.

While the CRM tools in the SAP ERP system, if employed properly, can help manage customer relationships, firms embracing the CRM concept often employ a separate CRM system that communicates with the ERP system. An advantage of this approach is that the planning and analysis performed in the CRM system do not interfere with the performance of the ERP system, which primarily processes large volumes of business transactions. SAP's ERP provides additional customer interaction functionality.

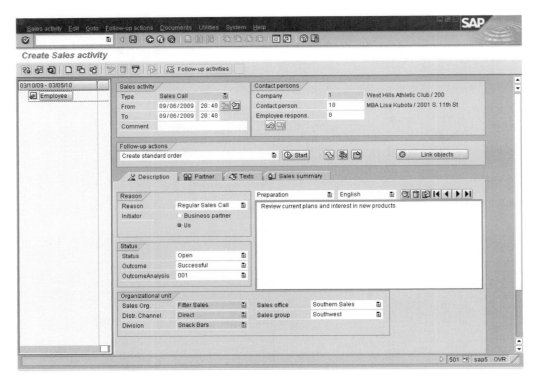

FIGURE 3-14 SAP ERP sales activity manager

Figure 3-15 shows how SAP CRM relates to the SAP ERP system as well as SAP's Business Warehouse (BW) and Advanced Planner and Optimizer (APO) modules.

As described previously, the SAP ERP system processes business transactions and provides much of the raw data for CRM. SAP's Business Warehouse is a flexible system for reporting and analysis of transactional data. By analyzing sales transactions using data mining, firms can discover trends and patterns to use in planning marketing activities. The Advanced Planner and Optimizer (APO) is a system that supports efficient planning of the supply chain. The APO's role in CRM is to provide higher levels of customer support through its Global Available-to-Promise (ATP) capabilities. In the standard SAP ERP system, the ATP capabilities function on a location-by-location basis. If the product or material a customer wants is not available in the location that usually serves the customer, then the sales order clerk can check for the material in other facilities, but this must be done on a facility-by-facility basis. With Global ATP, the system automatically checks all facilities and determines the most cost-efficient facility to use to meet the customer's request. The SAP CRM system communicates with the SAP ERP, BW, and APO systems in developing and executing its plans. Thus, CRM not only interfaces with the customer, but enables the company to analyze the customer data and best serve the customer.

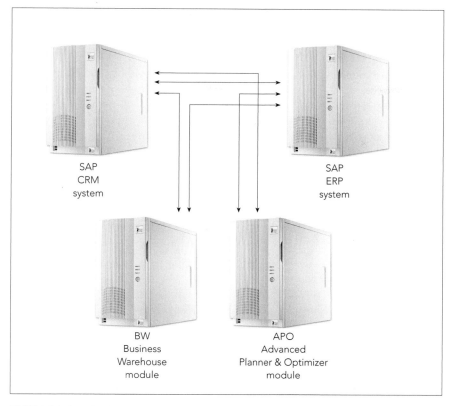

FIGURE 3-15   SAP CRM system landscape

SAP's view of CRM is to provide a set of tools to manage the three basic task areas, or jobs, related to customers: marketing, sales, and service. With CRM, these task areas contribute to the cultivation of the customer relationship. This cultivation goes through four phases, as defined by SAP: Prospecting, Acquiring, Servicing, and Retaining.

In Prospecting, a potential new customer (or potential new business with existing customers) is evaluated, and development activities (e-mails, sales calls, mailings, etc.) are planned to develop the prospective business. Marketing tasks predominate in this phase.

In Acquiring, salespeople develop business prospects into customers. Marketing is still the critical task, but the sales tasks (processing inquiries, quotes, and eventually sales orders) become increasingly important in this phase.

Once sales are established with the customer, the business becomes one of servicing the account. Sales tasks are still important, but service tasks (including technical support, warranty work, product returns, fixing quality problems, and complaint handling) are critical to maintaining customer satisfaction.

The rate at which prospects become customers is quite low; thus, a critical part of the process is Retaining. It is much easier to retain a good customer than to find a new one, so the focus of Retaining is making sure that current customers are satisfied by timely delivery of quality products and services at a fair price. Sales and service tasks are still critical, but marketing tasks are again important to anticipate changes in customers' requirements.

The customer development cycle (Prospecting, Acquiring, Servicing, and Retaining) is supported by Contact Channels, the methods the company uses to communicate with its customers. An interaction center provides contact through a variety of media (phone, fax, and e-mail). The Internet and mobile technologies are providing an increasingly large percentage of customer contacts. For example, the Contact Channels might aid a customer service agent on the telephone and prompt her to ask the customer various questions relating to the account.

Another set of tools in SAP CRM is Marketing and Campaign Management. Companies invest significant sums of money in marketing campaigns, which are promotional activities that publicize the product and the company. Successful planning, execution, and evaluation are critical to gain maximum effect from these efforts. Figure 3-16 shows how SAP CRM supports marketing and campaign management. The top half of this diagram represents planning activities, while the bottom half represents execution and evaluation activities. These activities are supported by most CRM products.

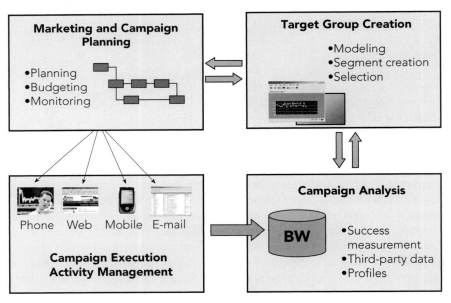

FIGURE 3-16   Marketing and campaign planning

Marketing and campaign planning includes task scheduling, resource allocation, and budgeting. These planning tasks are executed in conjunction with target group creation tasks, which use data from the SAP ERP system (perhaps using BW) to categorize the company's customers, offering them more individual product and service promotions.

Campaign Execution Activity Management is a set of tools to help manage the execution of the marketing campaign, which can include handling sales calls, mailings, personalized e-mailings, and Web-based promotional activities directed to the targeted group of customers. These activities can be monitored throughout the campaign to make sure they are completed. For example, the status of a planned customer phone call will remain open until the call is completed.

The Campaign Analysis tool allows the company to evaluate the success of the campaign so that it can incorporate improvements in the next marketing campaign. Marketing employees can use a number of measures to determine the success of the campaign, including tracking lead generation and response rates. Staff can use BW tools, by running queries, to support this analysis.

The Benefits of CRM

CRM provides companies with these benefits:

- Lower costs: CRM can lead to operational efficiencies, such as better response times in call center operations and better use of sales force time, which lower costs.
- Higher revenue: Segmenting customers leads to better selling opportunities and revenue increases.
- Improved strategy and performance measurement: Installing and operating an ERP system requires management and staff to think of the company as a whole. This attitude carries over into CRM work. With CRM in place, management can think about different performance measures; for example, should salespeople be rewarded for exceeding sales quotas, and marketing people rewarded for finding new customers—or, should both receive rewards that are based on some measure of customer satisfaction? The former approach, typical in days before CRM and ERP, can lead to unintegrated functional thinking. The latter approach—now feasible with CRM and ERP—can lead all personnel to think in terms of a company-wide effort to satisfy customers.

ANOTHER LOOK

CRM: Strategy and Demand

CIO magazine offers a tutorial on CRM that addresses many practical questions and considerations. For example, companies should be emphasizing CRM as a strategy, not just a technical solution. This approach, in many ways, is similar to the way companies should implement ERP solutions; if management just considers an ERP system to be a technical solution, without thinking about strategy and change management, the project is doomed to fail. CRM's strategy helps an organization understand its customers and grasp how to meet their requirements. This strategy translates into selling customers what they want, cross-selling if possible, obtaining new customers while retaining old ones, closing deals faster, and in general, offering better customer service. Companies can implement this CRM strategy through call centers, Web sites, advertising, or other channels. Patterns of customer behavior can be tracked from each of these areas and combined into a single depiction of the customer. According to *CIO*, if someone has multiple accounts with one bank, it is to the bank's benefit to treat this person well each time it has any contact with him or her, even if the employee serving that customer has very little business with him or her.

A CRM project should be run across all departments, like an ERP project. And management buy-in and commitment is critical for it to be successful. Traditionally, financial services and telecommunications organizations have been the first to adopt CRM. Manufacturing organizations are the last.

There has been a shift toward on-demand CRM, but some companies have reported problems with this newer delivery of the software. In 1999, Salesforce.com introduced on-demand CRM, which was an attractive option for small to midsized companies that wanted to get into CRM without a huge initial investment. However, integration can be tricky, especially with larger and more complex integration spanning many departments; upgrades are problematic; and privacy-sensitive organizations, such as health care, are reluctant to give up data to a third party.

Questions:

1. What are the advantages and disadvantages of on-demand CRM for a small to midsized company? What are the advantages and disadvantages for a large company?

2. On the Internet, visit Salesforce.com and report on the various products the company offers. Assume you are running a small Internet business that sells tickets to concerts and sporting events. Would you be able to use Salesforce software? Why, or why not?

Chapter Summary

- Fitter Snacker's unintegrated information systems are at the root of an inefficient and costly sales order process. Because information is not shared in real time, customers are asked to repeat initial sales order information. As an order is processed, errors in pricing, credit checks, and invoicing also occur, presenting a poor company image to customers. Integrated ERP software would let FS avoid errors because the system stores all customer data in a central database that is shared in real time by all company employees.

- An ERP system such as SAP ERP treats a sale as a sequence of related functions, including taking orders, setting prices, checking product availability, checking the customer's credit line, arranging for delivery, billing the customer, and collecting payment. In SAP ERP, all these transactions, or documents, are electronically linked, so tracking an order's status (partial shipments, returns, partial payments, and so forth) is easily accomplished.

- Installing an ERP system means making various configuration decisions, which reflect management's view of how transactions should be recorded and later used for decision making. For example, the system can be configured to limit selling price discounts, thus avoiding unprofitable pricing.

- An ERP system's central database contains tables of master data—relatively permanent data about customers, suppliers, material, and inventory—as well as transaction data tables, which store relatively temporary data such as sales orders and invoices.

- Customer relationship management (CRM) systems build on the organizational value ERP provides; they specifically increase the flexibility of the company's common database regarding customer service. Various kinds of CRM software are available, some from ERP vendors (including SAP) and some from third-party software companies. CRM software can lead to operational savings, but most companies buy it because they feel that having better customer relationships will result in higher revenues. Uses of CRM have evolved since the software was initially launched; what began as a customer contact repository has extended its capabilities to include sophisticated business intelligence. CRM can be installed in-house or on-demand.

Key Terms

Audit trail	Document flow
Condition technique	Material master data
Customer master data	On-demand
Customer relationship management (CRM) software	Organizational structure
Delivery	

Exercises

1. Assume you are the Fitter Snacker salesperson calling on the local headquarters for a chain of convenience stores in your area. You just started this job, and you are nervous about meeting your customer for the first time. Describe all the problems you might encounter when taking and filling the order, if you are using the old Fitter Snacker process described in this chapter. Now assume you have an ERP system in place. Describe the sequence of events involved in selling the snack bars to the convenience stores with the new system in place.

2. Assume Fitter Snacker plans to install an ERP system and reorganize its sales divisions, consolidating them into one. What personnel issues would the management have to deal with, when promoting the new system to the sales force? How would you deal with those issues?

3. Fitter Snacker's current sales order accounting involves recording sales in each sales division and then periodically sending enough data to Accounting to record sales for the company. Complete sales order data are retained in each sales division for business analysis purposes. Assume that different divisions of the Delicious Foods Company buy NRG-A and NRG-B bars from each of FS's sales divisions. To complicate matters, some divisions of Delicious buy store-brand bars from FS. (Delicious owns convenience store outlets.) FS management wants to see an analysis of the overall relationship with Delicious Foods. FS's management thinks there may be opportunities to promote the relationship with Delicious, but they need to assess profitability before proceeding. They want to see what products each division sells to Delicious, how much is sold, and the terms. Assume that in FS's current system, all the required data are available only at the sales division level. What steps will be needed to pull this company-wide analysis together? (Review how each division sells its products and keeps its records.) Do you think a sales division manager will be enthusiastic about sharing all data with his or her counterpart in the other division? Do you think there might be some reluctance? Why?

4. Continuing the Delicious Foods example, now assume that FS has an SAP ERP system installed. Each sales division records sales in the same way. Sales records exist in real time and are kept in the company's common database. What steps will be needed to pull this company-wide analysis together? Do you expect that the divisions will meet the new system with enthusiasm or reluctance?

5. Assume you are the CIO for Fitter Snacker. You are frustrated with the lack of information flow between Marketing and Sales, Accounting, and Supply Chain. You need to convince top management to approve an ERP implementation project that will solve this problem. Write a persuasive memo to top management, convincing them of this need. Focus on the lack of information flow to and from Marketing and Sales.

6. Describe how FS's SAP system simplifies looking up customer numbers, setting a delivery date, and charging a unique price to a given customer. Include a discussion of master data.

7. What is document flow? Why is it important for auditors of a company?

8. A CIO of a major pharmaceutical company once stated that the reason the corporation used ERP systems could be summed up in one word: *control*. How does an ERP system give management control?

9. How can a business better serve its customers using the APO tool in SAP?

10. Assume you are the marketing manager for a large consumer products company, such as Procter & Gamble. You need to launch a new marketing campaign. How can you convince your company of the value of using CRM to help with this campaign? What type of CRM system would you choose for your company?

For Further Study and Research

Database Systems Corp. White Paper. "The CRM Journey." 2007. http://www. databasesystemscorp.com/tech-crm_26.htm.

Deck, Stewart. "Crunch Time." *CIO,* September 15, 2000. http://www.cio.com/archive/09152000_crunch.html.

Edwards, John. "What's Your Problem?" *CIO,* September 1, 2000. http://www.cio.com/archive/090100_problem.html.

Fickel, Louise. "Know Your Customer." *CIO,* August 15, 1999. http://www.cio.com/archive/081599_customer.html.

Goff, John. "Head Games." *CFO,* July 1, 2004. http://www.cfo.com/printable/article.cfm/3014815/c_3046615?f=options.

Hildebrand, Carol. "One to a Customer." *CIO,* October 15, 1999. http://www.cio.com/archive/enterprise/101599_customer.html.

Kawamoto, Dawn and Margaret Kane. "Oracle to swallow Siebel for $5.8 billion." *CNET News.com,* September 12, 2005. http://news.com.com/Oracle+to+swallow+Siebel+for+5.8+billion/2100-1014_3-5860113.html?tag=item.

Kontzer, Tony. "Better Late Than Never? SAP Spices Up On-Demand CRM." *Information Week,* February 6, 2006. http://www.informationweek.com/showArticle.jhtml;jsessionid=C2AW4U2FFP1FMQSNDLRSKHSCJUNN2JVN?articleID=178601884.

————. "Business Intelligence Can Give CRM A Boost." *Information Week,* June 18, 2004. http://www.informationweek.com/showArticle.jhtml;jsessionid=C2AW4U2FFP1FMQSNDLRSKHSCJUNN2JVN?articleID=22100805.

MacRae, Duncan. "Forrester Research praises SAP CRM software." *ITP Technology,* April 16, 2007. http://www.itp.net/news/489176?tmpl=print&print=1&page=.

Murphy, Chris. "IT Confidential: Executive Swaps and Expletives." *Information Week,* August 9, 2004. http://www.informationweek.com/showArticle.jhtml;jsessionid=GDL5S2OTJIDE2QSNDLPCKHSCJUNN2JVN?articleID=26806480.

Overby, Stephanie. "The Truth about On-Demand CRM." *CIO,* January 15, 2006. http://www.cio.com/article/16546/The_Truth_About_On_Demand_CRM/1.

Pender, Lee. "CRM from Scratch." *CIO, August* 15, 2000. http://www.cio.com/archive/081500_scratch.html.

Peppers, Don and Martha Rogers. "Customer Value." *CIO,* September 15, 1998.

Thompson, Bob and Francis Buttle. "CRM Must Be Linked to Value: An Interview with Francis Buttle." CRMGuru.com, June 17, 2004.

Varon, Elana. "Suite Returns." *CIO,* August 15, 2000.

Wailgum, Thomas. "ABC: An Introduction to CRM." *CIO,* 2007. http://www.cio.com/article/40295.

CHAPTER 4

PRODUCTION AND SUPPLY CHAIN MANAGEMENT INFORMATION SYSTEMS

LEARNING OBJECTIVES

After completing this chapter, you will be able to:

- Describe the steps in the production planning process of a high-volume manufacturer such as Fitter Snacker.
- Describe Fitter Snacker's production and materials management problems.
- Describe how a structured process for Supply Chain Management planning enhances efficiency and decision making.
- Describe how production planning data in an ERP system can be shared with suppliers to increase supply chain efficiency.

INTRODUCTION

In Chapter 2, you learned that Enterprise Resource Planning (ERP) has its roots in materials requirements planning (MRP). In fact, MRP is still a large part of today's ERP systems. In this chapter, we'll look at Supply Chain Management (SCM) in an ERP system. Fitter Snacker is part of a supply chain that starts with farmers growing oats and wheat and ends with a customer buying an NRG bar from a retail store. First, we will examine how Fitter Snacker manages its production activities, and then we will look at the broader concept of supply chain management.

In Chapter 3, we looked at Fitter Snacker's sales order process, and we assumed that FS had enough snack bars in its warehouse to fill a typical order. Like most unintegrated manufacturing operations, however, FS often has trouble scheduling production. Consequently, sometimes its warehouse is not adequately stocked, and customer orders cannot be filled in a timely fashion, leading to customer dissatisfaction and lost sales. In this chapter, you'll explore FS's SCM problems and learn how ERP can help fix them.

PRODUCTION OVERVIEW

To meet customer demand efficiently, Fitter Snacker must develop a forecast of customer demand, then develop a production schedule to meet the estimated demand. Developing a production plan is a complicated task, but the end result answers two simple questions:

- How many of each type of snack bar should we produce, and when?
- What quantities of raw materials should we order so we can meet that level of production, and when should they be ordered?

Developing a good production plan is just the first part of serving customers: Fitter Snacker must be able to execute the plan and make adjustments when customer demand does not meet the forecast. An ERP system is a good tool for developing and executing production plans because it integrates the SCM functions of production planning, purchasing, materials management/warehousing, quality management, and sales and accounting. To support even better planning of the supply chain, companies can connect ERP systems to supplier and customer information systems as well.

In this chapter, we will use spreadsheet examples to illustrate the logic that FS should be using to plan and schedule production of the bars. First, the production process of manufacturing the bars is explained. After demonstrating the planning and scheduling logic at each stage of the production planning process, using spreadsheets, we will show the SAP ERP screens that implement the logic in an ERP environment. Throughout the chapter, you will learn why using an integrated information system is superior to using unintegrated systems.

The goal of production planning is to schedule production economically, so that the company can ship goods to customers by the promised delivery dates in the most cost-efficient manner. There are three general approaches to production:

1. *Make-to-stock* items are made for inventory (the "stock") in anticipation of sales orders. Most consumer products (for example, cameras, canned corn, and books) are made this way.
2. *Make-to-order* items are produced to fill specific customer orders. Companies usually take this approach when producing items that are too expensive to keep in stock or items that are made or configured to customer

specifications. Examples of make-to-order items are airplanes and large industrial equipment.

3. *Assemble-to-order* items are produced using a combination of make-to-stock and make-to-order processes. The final product is assembled for a specific order from a selection of make-to-stock components. Personal computers are a typical assemble-to-order product.

Fitter Snacker's Manufacturing Process

Fitter Snacker uses make-to-stock production to produce its snack bars. The manufacturing process is illustrated in Figure 4-1.

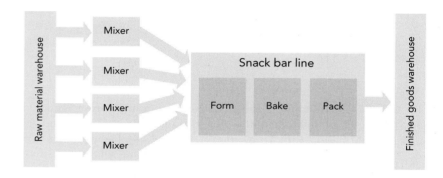

FIGURE 4-1 Fitter Snacker's manufacturing process

The snack bar line can produce 200 bars a minute, or 3,000 pounds of bars per hour. Each bar weighs 4 ounces. The entire production line operates on one shift a day. Here's how it works.

Fitter Snacker's Production Sequence

Raw materials are taken from the warehouse to one of four mixers. Each mixer mixes dough in 500-pound batches. Mixing a batch of dough requires 15 minutes of mixing time, plus 15 additional minutes to unload, clean, and load the mixer for the next batch of dough; therefore, each mixer can produce two 500-pound batches of dough per hour—more than the production line can process. Because only three mixers need to be operating at a time to produce 3,000 pounds of snack bars per hour, a mixer breakdown will not shut down the production line.

After mixing, the dough is dumped into a hopper (bin) at the beginning of the snack bar line. A forming mechanism molds the dough into bars, which will weigh 4 ounces each after baking. Next, an automated process takes the formed bars on a conveyor belt through an oven that bakes the bars for 30 minutes. When the bars emerge from the oven, they are individually packaged in a foil wrapper, and each group of 24 bars is packaged into a display box. At the end of the snack bar line, production personnel either stack the display boxes on pallets (for small orders) or pack the display boxes into shipping boxes and take them to the finished goods warehouse.

Changing the snack bar line from one type of bar to the other takes 30 minutes, for cleaning the equipment and changing the wrappers, display boxes, and shipping cases. Each night, a second shift of employees cleans all the equipment thoroughly and sets it up for the next day's production. Thus, changing production from NRG-A on one day to NRG-B the next day can be done at the end of the day without a loss of capacity. **Capacity** is the amount of bars that can be produced. On the other hand, producing two products in one day results in a half-hour loss of capacity during the changeover.

Fitter Snacker's Production Problems

Fitter Snacker has no problems *making* snack bars. Fitter Snacker has problems deciding *how many* bars to make and *when* to make them. The manufacturing process at FS suffers from a number of problems, from communication breakdowns and inventory issues to accounting inconsistencies, mainly stemming from the unintegrated nature of its information systems.

Communication Problems

Communication breakdowns are inherent in most companies, and they are magnified in a company with an unintegrated information system. For example, FS's Marketing and Sales personnel do not share information with Production personnel: M/S frequently excludes Production from meetings, doesn't consult Production when planning sales promotions, fails to notify Production of planned promotions, and gives Production no warning that it has taken an exceptionally large order.

When Production must meet an unexpected increase in demand, several things happen. First, an unexpected spike in sales depletes warehouse inventories. To compensate, Production must schedule overtime labor, which results in higher production costs for products. Second, some materials, such as ingredients, wrappers, and display boxes, are custom products purchased from a single vendor. A sudden increase in sales demand might cause shortages or even a stockout of these materials. Getting them to the plant might require expedited shipping, further increasing the cost of production. Finally, Production personnel are evaluated on their performance—how successful they are at controlling costs, keeping manufacturing lines running, maintaining quality control, and operating safely. If they can't keep production costs down, Production staff receive poor evaluations. Managers are especially frustrated when an instant need for overtime follows a period of low demand. With advance notice of a product promotion by Marketing, they could have used the slack period to build up inventory in anticipation of the increase in sales.

Inventory Problems

Fitter Snacker's week-to-week and day-to-day production planning is not linked to expected sales levels in a systematic way. When deciding how much to produce, the production manager applies rules developed through experience. Her primary indicator is the difference between the normal amount of finished goods inventory that should be stocked and the actual inventory levels of finished goods in the warehouse. Thus, if NRG-A or NRG-B inventory levels seem low, the production manager schedules more bars for production.

However, she doesn't want too many bars in inventory because they have a limited shelf life. Her judgment is also influenced by the information she hears informally from people in Marketing about expected sales.

The production manager's inventory data are in an Access database. Data records are not maintained in real time, and they do not flag inventory that has been sold but not yet shipped. Such inventory is not available for sale, of course, but employees cannot determine this from the records, and thus do not know the level of inventory that is available to ship. This is problematic if the Wholesale Division generates unusually large orders or high volumes of orders. For example, two large Wholesale Division orders arriving at the same time can deplete the entire available inventory of NRG-A bars. In this case, the Production department must change the production schedule for NRG-B bars so it can fill the orders for NRG-A. This changeover means delaying production of NRG-B bars, and losing production capacity due to the unplanned production changeover.

The production manager lacks not only a systematic method for meeting anticipated sales demand, but also a systematic method for adjusting production to reflect actual sales. M/S does not share actual sales data with the Production department, partly because this information is hard to gather on a timely basis and partly because of a lack of trust between the Sales and Production departments (as a result of prior negative experiences or competition between the departments). Hence, the production manager must use warehouse inventory levels as a benchmark; standard levels of inventory are the only guide available. If Production had access to sales forecasts and real-time sales order information, the manager could make timely adjustments to production, if needed. These adjustments would allow inventory levels to come much closer to what is actually needed.

Accounting and Purchasing Problems

Production and Accounting do not have a good way to accumulate the day-to-day costs of FS's production. As discussed in Chapter 3, the warehouse keeps a fairly good running count of what is on hand. Furthermore, the company takes a monthly inventory, and actual counts are usually within 5 percent of what is on the books. Management would like to be more accurate, of course, but without a real-time inventory system, FS cannot achieve higher inventory accuracy.

Manufacturing costs are based on the number of bars produced each day, a number that is measured at the end of the snack bar production line. Fitter Snacker uses standard costs for the purpose of figuring manufacturing costs. **Standard costs** are the normal costs of manufacturing a product, and they are calculated from historical data and any changes in manufacturing that have occurred since the collection of the historical data. For each batch of bars it produces, FS can estimate direct costs (materials and labor) and indirect costs (factory overhead). The number of batches produced is multiplied by the standard cost of a batch, and the resulting amount is charged to manufacturing costs.

Most manufacturing companies use standard costs in some way, but the method requires that standards be adjusted periodically to conform with actual costs. (These adjustments will be discussed in Chapter 5.) Actual Fitter Snacker raw material and labor costs often deviate from the standard costs. FS is not good at controlling raw materials purchases, and the production manager cannot give the Purchasing manager a good production forecast. So the Purchasing manager works on two tracks: First, she tries to keep inventories high to avoid stockouts. Second, if she's offered good bulk-quantity discounts on raw materials such as oats, she will buy in bulk, especially for items that have long lead times for delivery. These purchasing practices make it difficult to forecast the volume of raw materials that will be on hand and their average cost. FS also has trouble accurately forecasting the average cost of labor for a batch of bars because of the frequent need for overtime labor.

Thus, Production and Accounting must periodically compare standard costs with actual costs and then adjust the accounts for the inevitable differences. This is always a tedious and unpleasant job. The comparison should be done at each monthly closing, but FS often puts it off until the closing at the end of each quarter, when its financial backers require legitimate financial statements. The adjustments are often quite large, depending on production volumes and costs during the quarter.

Exercise 4.1

a. A convenience store chain offers to buy a very large amount of its "store brand" health bars (the NRG-B bars with a customized wrapper). The chain wants a lower-than-normal selling price, but the proposed order is quadruple the size of its regular order. The marketing manager asks managers from Production, Purchasing, and Accounting whether the terms of the proposed deal will be profitable. Why will the managers in these areas have trouble providing a reliable answer on short notice?

b. The production manager notes that warehouse inventory levels are fairly high, so the production line does not need to be run for a full eight hours each day during the coming week. She plans to run the line for eight hours a day anyway, because if the line were down, workers would still need to be paid during the idle time, and overhead costs would be incurred as well. Running the line full-time will also decrease the average cost of bars actually produced (indirect costs can be spread over more bars), and some warehoused raw materials will spoil if they're not used soon. Is the production manager's reasoning logical? Why, or why not?

THE PRODUCTION PLANNING PROCESS

In this section, you will examine a systematic process for developing a production plan that takes advantage of an ERP system. Production planning is a complicated process. Spreadsheet calculations are presented to explain and illustrate the key steps, and the corresponding screens in the SAP ERP system follow the spreadsheet data.

Production planners are employees who interact with the inventory system and the sales forecast to figure out how much to produce. They follow three important principles:

• Work from a sales forecast and current inventory levels to create an "aggregate" ("combined") production plan for all products. Aggregate production plans help to simplify the planning process in two ways: First, plans are made for groups of related products rather than for individual products. Second, the time increment used in planning is frequently a month or a quarter, while the production plans that will actually be executed operate on a daily or weekly basis. Aggregate plans should consider the available capacity in the facility.

- Break down the aggregate plan into more specific production plans for individual products and smaller time intervals.
- Use the production plan to determine raw material requirements.

Production planners aggregate products into product groups to reduce the number of variables they must consider when developing a production plan. Developing production groups can become complicated. For example, cereal manufacturers can group together different package sizes, or they might group together product brands (such as kids' cereals, health cereals, and so on). A consumer products company may group by product types (e.g., shampoo, laundry detergent, and disposable diapers). The aggregate production plan for Fitter Snacker will combine the only two products, NRG-A and NRG-B bars, into one group to illustrate the process. The plan will be developed using a monthly time increment; then, this monthly production plan will be disaggregated to determine weekly raw material orders and daily production schedules.

The SAP ERP Approach to Production Planning

The SAP ERP approach to the production planning process is shown in Figure 4-2. Refer to this figure throughout this section to trace production planning.

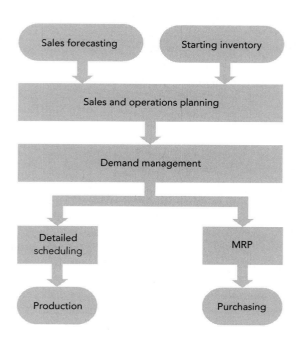

FIGURE 4-2   The production planning process

The information at each stage of the production process flows through the following steps. Each step is explained in detail in the next section of the chapter.

- *Sales forecasting* is the process of developing a prediction of future demand for a company's products.
- *Sales and operations planning (SOP)* is the process of determining what the company will produce. The Sales forecasting and Starting Inventory levels in the diagram are inputs to this process. At first glance, it would seem that a company should just make products to match the forecasted sales, but developing the production plan can be more complicated, because capacity must be considered. Many products have seasonal demand, and to meet demand during peak periods, production planners must decide whether to build up inventory levels before the peak demand, increase capacity during the peak period, subcontract production, or use some combination of these approaches.
- *Demand management* is the process of breaking down the production plan into finer time units, such as weekly or even daily production figures, to meet demand for individual products.
- *Detailed scheduling* uses demand management's production plans as an input for a production schedule. Methods of detailed scheduling depend on the manufacturing environment. For Fitter Snacker, the detailed production schedule will determine when to switch between the production of NRG-A and NRG-B bars.
- *Production* uses the detailed schedule to manage daily operations, answering the questions, "What should we be producing?" and "What staffing do we need to produce those products?"
- *Materials requirements planning (MRP)* determines the amount and timing of raw material orders. This process answers the questions, "What raw materials should we be ordering so we can meet a particular level of production?" and "When should we order these materials?"
- *Purchasing* takes the quantity and timing information from MRP and creates purchase orders for raw materials, which it transmits to qualified suppliers.

Let's take a more detailed look at each of these steps in the production process.

Sales Forecasting

A range of forecasting techniques can be used to predict consumer demand. Fitter Snacker has no formal way of developing a sales forecast and sharing it with Production. SAP's ERP system takes an integrated approach. Whenever a sale is recorded in the Sales and Distribution (SD) module, the quantity sold is recorded as a consumption value for that material. These consumption values can be maintained on a weekly or monthly basis, as desired. If more detail is needed, the Logistics Information system that is part of SAP ERP can record sales with more detail (for example, by region or sales office), or data can be stored in the separate Business Warehouse (BW) system for more detailed analysis. With an integrated information system, accurate historical sales data are available for forecasting.

One simple forecasting technique is to use a prior period's sales and then adjust those figures for current conditions. To make a forecast for Fitter Snacker, we can use the previous year's sales data in combination with marketing initiatives to increase sales. Look at the forecasts shown in Figure 4-3.

Sales forecasting		Jan.	Feb.	March	April	May	June
Previous year (cases)		5734	5823	5884	6134	6587	6735
Promotion sales (cases)						300	300
Previous year base (cases)		5734	5823	5884	6134	6287	6435
Growth:	3.0%	172	175	177	184	189	193
Base projection (cases)		5906	5998	6061	6318	6476	6628
Promotion (cases)							500
Sales forecast (cases)		5906	5998	6061	6318	6476	7128

FIGURE 4-3   Fitter Snacker's sales forecast for January through June

The sales data in Figure 4-3 are for shipping cases, which contain 12 display boxes that contain 24 bars each, for a total of 288 bars. Note in Figure 4-3 that the forecast starts with the previous year's sales levels, to reflect FS's seasonal sales fluctuations (sales are higher in the summer when more people are active). Also note that there was a special marketing promotion last year. The estimated impact of this promotion was an increase in sales of 300 cases for May and June. This increase must be subtracted from the previous year's sales values to get an accurate base measurement. FS's Marketing department anticipates a 3 percent growth in sales over the previous year, based on research reported in trade publications. And finally, FS will be launching a special marketing promotion at the end of May to increase sales at the beginning of the summer season. FS marketing experts think this will result in an increase in sales of 500 cases for June.

Exercise 4.2

Following the format of the spreadsheet shown in Figure 4-3, develop a spreadsheet to forecast Fitter Snacker's sales for July through December. Make the sales growth rate of 3 percent an input value, and calculate the base projection using the previous year's values, shown in Figure 4-4. Assume that the special marketing promotion last year resulted in an increase in sales of 200 cases for July, and that a special marketing promotion this year will result in an increase in sales for July of 400 cases.

Sales volume	July	Aug.	Sept.	Oct.	Nov.	Dec.
Previous year	6702	6327	6215	6007	5954	5813

FIGURE 4-4   Fitter Snacker's sales for the previous period, July through December

Sales and Operations Planning

Sales and operations planning (SOP) is the next step in the production planning process. The input to this step is the sales forecast provided by Marketing. The output of this step is a production plan designed to balance market demand with production capacity. This production plan is the input to the next step, demand management.

A sales and operations plan is developed from a sales forecast and determines how Manufacturing can efficiently produce enough goods to meet projected sales. In Fitter Snacker's case, there is no way to make this determination, because FS does not produce a formal estimate of sales. If FS had an ERP system, the calculation would be done as described here.

We know that Fitter Snacker can produce 200 bars per minute, so we can estimate the production capacity required by the sales forecast. Figure 4-5 shows FS's sales and operations plan for the first six months of the year.

Sales and operations planning		Dec.	Jan.	Feb.	March	April	May	June
1) Sales forecast			5906	5998	6061	6318	6476	7128
2) Production plan			5906	5998	6061	6318	6900	6700
3) Inventory		100	100	100	100	100	524	96
4) Working days			22	20	22	21	23	21
5) Capacity (shipping cases)			7333	6667	7333	7000	7667	7000
6) Utilization			81%	90%	83%	90%	90%	96%
7) NRG-A (cases)	70.0%		4134	4199	4243	4423	4830	4690
8) NRG-B (cases)	30.0%		1772	1799	1818	1895	2070	2010

FIGURE 4-5   Fitter Snacker's sales and operations plan for January through June

At the start of January, the production planner is projecting a beginning inventory of 100 cases. The first line in Figure 4-5 is the sales forecast, which is the output of the sales forecasting process shown in Figure 4-3. The next line is the production plan, which is developed by the production planner in a trial-and-error fashion, observing the effect of different production quantities on inventory levels and capacity utilization (the amount of plant capacity that is being consumed). The goal is to develop a production plan that meets demand without exceeding capacity and that maintains "reasonable" inventory levels (neither too high or two low). This process requires judgment and experience. The third line, inventory, is the difference between the sales forecast and the production plan. The production planner has developed a plan that maintains a minimum planned inventory of 100 cases. This inventory, called safety stock, is planned so that if sales demand exceeds the forecast, sales can be met without altering the production plan. Notice that in May, the production plan is greater than the May sales forecast, and the inventory is 524. Why? Because the planner wants to build up inventory to handle the increased demand in June, which results from the normal seasonal increase in snack bar sales and additional demand from the planned promotional activities. The fourth line is working days, an input based on the company calendar. Using the number of working days in a month, the available capacity each month is calculated in terms of the number of shipping cases.

200 bars per minute × 60 minutes per hour × 8 hours per day = 96,000 bars per day, which equals 333.3 cases per day (96,000 bars per day ÷ 24 bars per box ÷ 12 boxes per case).

If you multiply the number of working days in a month by the production capacity of 333.3 shipping cases per day, you get the monthly capacity in shipping cases, which is shown in line 5.

With the available capacity (assuming no overtime) now expressed in terms of shipping cases, it is possible to determine the capacity utilization for each month by dividing the production plan amount (line 2) by the available Capacity (line 5) and expressing the result as a Utilization percentage (line 6). This capacity calculation shows whether FS has the capacity necessary to meet the production plan. While higher levels of capacity utilization mean that Fitter Snacker is producing more with its production resources, this percentage must be kept below 100 percent to allow for production losses due to product changeovers, equipment breakdowns, and other unexpected production problems.

The last step in sales and operations planning is to disaggregate the plan, that is, break it down into plans for individual products. Lines 7 and 8 disaggregate the planned production shown in line 2, based on the breakdown of 70 percent NRG-A and 30 percent NRG-B snack bars. This 70/30 breakdown is based on previous sales data for these products. These monthly production quantities are the output of the sales and operations planning process, and are the primary input to the demand management process.

Suppose that FS is regularly able to achieve production levels at 90 percent of capacity. If the sales forecast requires more than 90 percent capacity, FS management can choose from among the following alternatives to develop a production plan:

1. FS might choose not to meet all the forecasted sales demand, or it might reduce promotional activities to decrease sales.
2. To increase capacity, FS might plan to use overtime production. Doing that, however, would increase labor cost per unit.
3. Inventory levels could be built up in earlier months, when sales levels are lower, to reduce the capacity requirements in later months. Doing that, however, would increase inventory holding costs and increase the risk that NRG bars held in inventory might pass their expiration date before being sold by retailers.
4. To find the right balance, management might try a hybrid approach to the capacity problem: reduce sales promotions slightly, increase production in earlier months, and plan for some overtime production.

The monthly production quantities in lines 7 and 8 of Figure 4-5 create some inventory in May to meet June's sales; in addition, some overtime production is likely in June because capacity utilization is over 90 percent. This example illustrates the value of an integrated system: it provides a tool to incorporate data from Marketing and Manufacturing to evaluate different plans. Whereas Marketing may want to increase sales, the company might not increase its profits if overtime costs or inventory holding costs are too high. This sort of planning is difficult to do without an integrated information system, even for small companies like FS. Having an integrated information system helps managers of all functional areas meet corporate profit goals.

In SAP ERP, the sales forecast can incorporate historical sales data from the Sales and Distribution (SD) module, or input from plans developed in the Controlling (CO) module can form the basis for the forecast. In the CO module, profit goals for the company can be set, which can then be used to estimate the sales levels needed to meet the profit goals. Figure 4-6 (on the next page) shows the sales and operations planning screen from the SAP ERP system. The title of this screen is "Create Rough-Cut Plan." **Rough-cut planning** is a common term in manufacturing for aggregate planning. As described above, rough-cut plans are disaggregated to generate detailed production schedules.

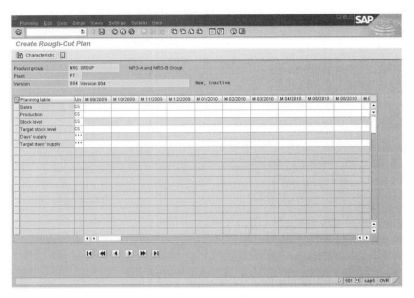

FIGURE 4-6   Sales and operations planning screen in SAP ERP

The sales forecast is entered in the first row (Sales) of the rough-cut plan. Data can be entered manually by the user, a sales forecast can be transferred from a profitability analysis performed in the CO module, or the user can perform a forecast in this screen, calling up historical sales data from the SAP ERP system. The second row, Production, represents the production that is planned to meet the sales forecast. It, too, can be entered manually, or the SAP system can generate values that meet sales goals. The third row shows inventory as the Stock level. The gray color indicates that it is a calculated result. The fourth row allows for the entry of a Target stock level. If the user enters a value in this row, then SAP ERP will propose production levels to meet the Target stock level. Once the Target stock level is entered in the fourth row, the system will calculate the number of Days' supply, so the fifth row is a calculated result. The sixth row lets the user specify a Target stock level in terms of the number of days of demand it would cover, known as Target days' supply. The SAP system uses the factory calendar, which specifies company holidays and planned shutdowns, to determine the number of working days in a month when calculating the Target days' supply.

If the sales plan (first line in Figure 4-6) is to be developed using forecasting tools, the SAP ERP system will provide the planner with historical sales values based on sales data stored in the system. Without an integrated system like SAP, the planner would likely have to request sales figures from the Sales department, and might not be sure how accurate the data were. Figure 4-7 shows how the SAP ERP system displays historical sales figures. In addition to providing the historical sales values from the SD module, this screen allows the planner to "correct" the sales values. For example, sales may have been low in the past due to unusual weather conditions, or the planner might know that sales would have been higher if the company could have met all demand. The sales figures used for forecasting should represent the best estimate of what *demand* was in the past, not necessarily what the actual sales were.

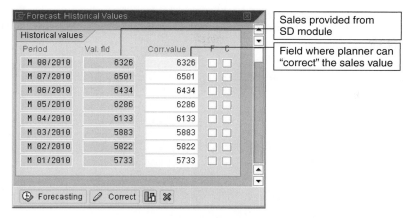

FIGURE 4-7   Historical sales levels for Fitter Snacker

The SAP ERP system can automatically graph these data to help the planner determine if there are any unusual patterns in the historical sales values that require investigation. The planner can "correct" these values as well, to adjust sales values that were unusually high or low, or to "back out" the effects of previous sales promotions. After the sales forecast is made, it can be adjusted to incorporate increased sales from planned sales promotions. Once the historical data are acceptable, the user can select one of the SAP ERP tools shown in Figure 4-8 to prepare the forecast.

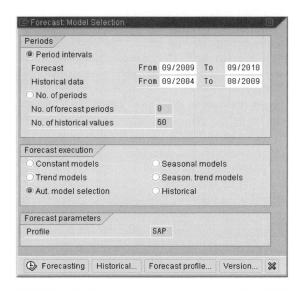

FIGURE 4-8   Forecasting model options in SAP ERP

This screen allows the user to specify a number of forecasting parameters, including whether the model should allow for trends and seasonal variations. Once the SAP ERP system generates a forecast, the planner can view the results graphically, as shown in Figure 4-9. While the SAP ERP system also provides the standard statistical measures of forecast accuracy, human judgment is frequently the best determinant of whether the forecast results make sense.

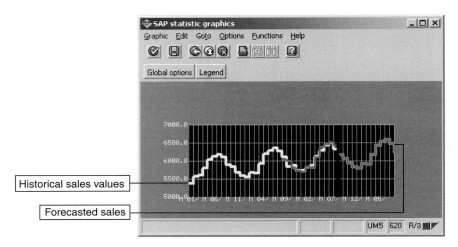

FIGURE 4-9   Forecasting results presented graphically in SAP ERP

Another feature of the SAP ERP sales and operations planning process is the integration of rough-cut capacity planning, which is shown in Figure 4-10. Rough-cut capacity planning applies simple capacity-estimating techniques (like those shown in the spreadsheet example in Figure 4-5) to the production plan, to see if the techniques are feasible. Frequently, rough-cut capacity planning techniques are applied to critical resources—those machines or production lines where capacity is usually limited. For a company like Fitter Snacker, with a simple manufacturing process, these estimates can be very accurate. For more complex manufacturing processes, these estimates will not be completely accurate but will ensure that the production plans are at least reasonable. Managers can use SAP ERP's more sophisticated planning tools at the detailed scheduling level, when the plans that are developed will actually be converted into manufacturing decisions on the shop floor.

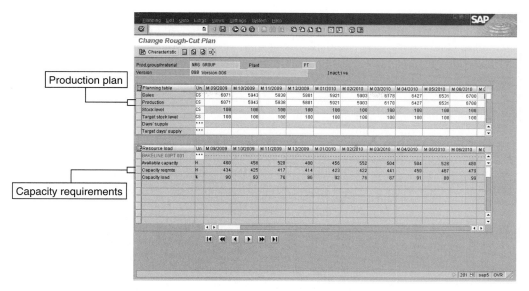

Production plan

Capacity requirements

FIGURE 4-10   Sales and operation plan with rough-cut capacity calculation in SAP ERP

While an integrated ERP system like SAP can provide sophisticated tools to support the sales and operations planning process, the plan will only be successful if the interested parties are committed to the process. If Marketing and Manufacturing cannot agree on sales forecasts, promotions, and production plans, then the company will find that it is over-stocked with some items, running out of others, and spending a lot of money on over-time production and expedited shipping. Successful sales and operations planning depend on developing a culture of cooperation between Marketing and Manufacturing, which usually requires continuous support, encouragement, and enforcement from top management. As you will learn in Chapter 7, changing a company's culture is usually a much harder challenge than the technical challenge of installing new hardware and software.

ANOTHER LOOK

Kellogg Company: Sales and Operations Planning

The Kellogg Company has realized tremendous savings and efficiencies from a coordinated sales and operations planning process.

A key to the success of this effort was changing the focus of the key players. Previous to the new process, Marketing and Sales personnel were evaluated on how many tons of cereal they sold, and Manufacturing was evaluated on how many tons of cereal they produced. Nobody focused on whether they were making a profit. The Marketing area was

continued

focusing on national sales promotions, which created huge peaks in demand. Sales personnel were making sales deals that moved a lot of cereal, but at terms that didn't necessarily make profits for the company. Sales forecasts were often incorrect, and last-minute customer orders created inefficient distribution and logistics practices as products were often redeployed to fulfill these rush orders. Production personnel focused on making as much cereal as possible, which meant that they minimized product changeovers so as not to lose production throughput and line capacity during the changeovers. These long production runs meant that a lot of cereal was being produced—but it wasn't necessarily the cereal that was in demand by customers. With a focus on minimizing waste in the plants, cereal was often produced to avoid scrapping packaging, even though the demand forecast did not require the additional production. This overproduction and misdeployed production resulted in old and damaged product that needed to be liquidated or written off as unsalable goods.

The Kellogg Company now has a structured sales and operations planning process called Integrated Business Planning (IBP), with all participants focused on making profitable decisions for the company. Marketing has moved from national product promotions to customer-specific product promotions that are time-phased to reduce the up-and-down demand cycle and inventory peaks. Because the plans are coordinated with Manufacturing, the production schedules are developed to support the promotions and to produce to the forecast with a minimum of safety stock. As Manufacturing is now evaluated on whether the company makes money, it plans production so that it produces the products that are in the demand forecast and can be sold. The profit goal motivates the sales force to negotiate deals that make money for the company, not deals at any cost merely to sell cereal. There is also an overall focus on value (dollar revenue) versus volume (production tonnage).

Developing an effective sales and operations planning process at the Kellogg Company was not easy, but the rewards have been significant. The process is an important factor in the recent success of the company. In October 2004, the Kellogg Company Web site reported third-quarter earnings growth driven by increased sales, investment, and effective execution. The company raised its earnings guidance for the full year of 2004 due to these strong results and continued business momentum. Because Marketing, Sales, and Manufacturing are working together, along with Finance, the Kellogg Company has been able to reduce its production capacity, finished goods inventory, and capital requirements, while selling more cereal than ever before, because it is producing the right product in the right quantities at the right time.

The results have manifested themselves in fresher products and improved order fulfillment for the consumer, fewer stockouts for retailers, reduced damages and unsalable merchandise, lower freight and warehouse costs, more efficient customer promotion investments, and significantly increased cash flow. These improvements combine to increase revenues, profits, share of market, and shareholder equity.

Question:

1. Write a memo to the CIO, manufacturing manager, and marketing manager of Fitter Snacker, convincing them of the importance of a sales and operations planning process. Use the Kellogg Company example to be persuasive.

Disaggregating the Sales and Operations Plan

As mentioned previously, companies typically develop sales and operations plans for product groups. Product groups are especially important for companies that have hundreds of products, because developing unique plans for hundreds of individual products is extremely time-consuming. Furthermore, it would be hard to develop these plans in a coordinated fashion so that production capacity could be considered. Fitter Snacker's product group is very simple, consisting of 70 percent NRG-A and 30 percent NRG-B bars. Figure 4-11 shows how product groups are defined in the SAP ERP system. The system allows any number of products to be assigned to a product group. Product groups can have other product groups as members, as well, so complicated aggregations can be defined.

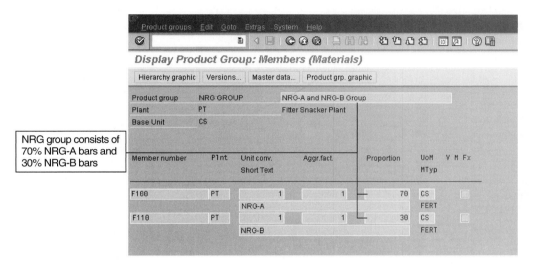

NRG group consists of 70% NRG-A bars and 30% NRG-B bars

FIGURE 4-11 Product group structure in SAP ERP

When the sales and operation plan is disaggregated, the production plan quantities specified for the group are transferred to the individual products that make up the group, according to the percentages defined in the product group structure (Figure 4-11).

The results of the disaggregation process can be seen in the Stock/Requirements List shown in Figure 4-12 (on the next page). This screen displays the inventory level for an individual product, including all planned additions and reductions. As you can see in Figure 4-12, the production plan from the sales and operation plan has been added to the Stock/Requirements List for NRG-A bars as reductions to the inventory levels.

The Stock/Requirements List is aptly named. It shows current stock, required materials, material expected to be received, and availability.

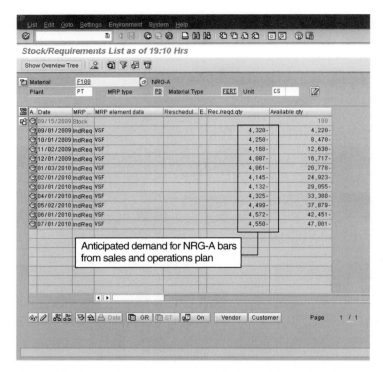

FIGURE 4-12 Stock/Requirements List for NRG-A bars after disaggregation

Exercise 4.3

Using the sales forecast for July through December from Exercise 4.2, develop a spreadsheet for sales and operations planning. Use the format of the spreadsheet shown in Figure 4-5. The number of working days for each month is shown in Figure 4-13.

	July	Aug.	Sept.	Oct.	Nov.	Dec.
Working days	22	18	20	23	20	17

FIGURE 4-13 The number of working days at Fitter Snacker, July through December

Note that there are fewer working days in August because of an annual plant shutdown. The number of working days is also low in December because of the Christmas and New Year holidays.

For your production plan, try to keep the capacity utilization at 95 percent or less. To disaggregate the plan for the group into plans for NRG-A and NRG-B bars, use 70 percent of sales for NRG-A bars and 30 percent for NRG-B bars.

Demand Management

The demand management step of the production planning process links the sales and operations planning process with the detailed scheduling and materials requirements planning processes. The output of the demand management process is the **master production schedule (MPS)**, which is the production plan for all finished goods. For Fitter Snacker, the MPS is an input to detailed scheduling, which determines which bars to make and when to make them. The MPS is also an input to the MRP process, which determines what raw materials to order to support the production schedule.

The demand management process splits FS's monthly production planning values into finer time periods. Figure 4-14 shows January's production plan by week and by day.

Demand management		Week 1	Week 2	Week 3	Week 4	Week 5	
		1/2 - 1/5	1/8 - 1/12	1/15 - 1/19	1/22 - 1/26	1/29 - 1/31	2/1 - 2/2
Monthly demand	NRG-A	4134	4134	4134	4134	4134	4198
	NRG-B	1772	1772	1772	1772	1772	1799
Working days in week		4	5	5	5	3	2
Working days in month		22	22	22	22	22	20
MPS	NRG-A	752	940	940	940	984	
Weekly demand	NRG-B	322	403	403	403	422	

Demand management		Jan 2	Jan 3	Jan 4	Jan 5	Jan 6
	NRG-A	4134	4134	4134	4134	4134
	NRG-B	1772	1772	1772	1772	1772
Working days in month		22	22	22	22	22
MPS	NRG-A	188	188	188	188	188
Daily demand	NRG-B	81	81	81	81	81

FIGURE 4-14 Fitter Snacker's production plan for January: The first five weeks of production are followed by a day-by-day disaggregation of week 1

FS will use the weekly plan to plan materials management for purchasing. Daily plans will be used for the product(s) that are to be produced on the snack bar line. The calculations were performed as follows:

- For the weekly plan, the MPS plan for NRG-A bars in week 1 was calculated as: 4,134 cases in Jan. (monthly demand) × 4 working days in week 1 ÷ 22 working days in month of Jan. = 751.6 cases

This figure was rounded to 752 cases in Figure 4-14.

- Because week 5 consists of the last three days in January and the first two days in February, the MPS for NRG-A bars in week 5 was calculated as:

4,134 cases in Jan. (monthly demand) × 3 working days in week 5 ÷ 22 working days in month of Jan. = 563.7 cases

4,198 cases in Feb. (monthly demand) × 2 working days in week 5 ÷ 20 working days in month of Feb. = 419.8 cases

Total = 563.7 + 419.8 = 983.5 cases

Master production schedule cases for week 5 were rounded to 984. In Figure 4-14, the daily production plan values for NRG-A and NRG-B bars were calculated by taking the monthly production plan value and dividing it by the number of working days in the month.

Notice that the demand management process for Fitter Snacker involves no user input. Input is from the sales and operations planning step. The SAP ERP software also uses information from the factory calendar (working days in week and month) to calculate the MPS.

Fitter Snacker does not do this sort of planning because it has no way to formally share sales forecast data between Marketing and Production. The company cannot relate its possible sales to its capacity and to the time available to make the product. Thus, FS cannot produce an accurate master production schedule.

The MPS is an input to the detailed scheduling and MRP processes. MRP is discussed next, and detailed scheduling is discussed after that.

Exercise 4.4

Develop a weekly production plan for July, like the one for January shown in Figure 4-14. For the weekly sales periods, the last week will include three days in August. The factory calendar information is shown in Figure 4-15.

	Week 1	Week 2	Week 3	Week 4	Week 5	
Demand management	7/2 - 7/6	7/9 - 7/13	7/16 - 7/20	7/23 - 7/27	7/30 - 7/31	8/1 - 8/3
Working days in week	4	5	5	5	2	3
Working days in month	22	22	22	22	22	18

FIGURE 4-15   Fitter Snacker's factory calendar for July

Materials Requirements Planning (MRP)

Materials requirements planning (MRP) is the process that determines the required quantity and timing of the production or purchase of subassemblies and raw materials needed to support the MPS. The MRP process answers the questions, "What quantities of raw materials should we order so we can meet that level of production?" and "When should these materials be ordered?" In this section, you will see examples of how Fitter Snacker could accurately plan its raw materials purchases if it had an ERP system.

In Fitter Snacker's case, all product components (ingredients, snack bar wrappers, and display boxes) are purchased, so the company could use the MRP process to determine the timing and quantities for purchase orders. To understand MRP, you must understand the bill of material, the material's lead time, and the material's lot sizing.

Bill of Material

The **bill of material (BOM)** is a list of the materials (including quantities) needed to make a product. The BOM for a 500-pound batch of the NRG-A or NRG-B bars is shown in Figure 4-16.

Ingredient	Quantity	
	NRG-A	NRG-B
Oats (lb.)	300	250
Wheat germ (lb.)	50	50
Cinnamon (lb.)	5	5
Nutmeg (lb.)	2	2
Cloves (lb.)	1	1
Honey (gal.)	10	10
Canola oil (gal.)	7	7
Vit./min. powder (lb.)	5	5
Carob chips (lb.)	50	
Raisins (lb.)	50	
Protein powder (lb.)		50
Hazelnuts (lb.)		30
Dates (lb.)		70

FIGURE 4-16   The bill of material (BOM) for Fitter Snacker's NRG bars

The BOM for Fitter Snacker's NRG bars is fairly simple because all ingredients are mixed together to form the dough; there are no intermediary steps. To produce many other products, however, component parts are joined into subassemblies that are then joined to form the finished product. It is obviously more complicated to calculate the raw material requirements for products with more complex BOMs.

Lead Times and Lot Sizing

The BOM can be used to calculate how *much* of each raw material is required to produce a finished product. Determining the *timing* and *quantity* of purchase orders, however, requires information on lead times and lot sizing.

For example, if a manufacturer orders a make-to-stock item, the **lead time** is the cumulative time required for the supplier to receive and process the order, take the material out of stock, package it, load it on a truck, and deliver it to the manufacturer. The manufacturer might also include the time required to receive the material in its warehouse (unloading the truck, inspecting the goods, and moving the goods into a storage location).

Lot sizing refers to the process of determining production quantities (for raw materials produced in-house) and order quantities (for purchased items). In FS's case, many raw materials can only be ordered from a supplier in certain bulk quantities. For example, because FS uses large quantities of oats, the most cost-effective way to purchase oats is in bulk hopper-truck quantities, which means that the material must be ordered in 44,000-pound quantities. Wheat germ, however, is used in smaller quantities, and to avoid having wheat germ become stale, FS orders it in 2,000-pound bulk containers. Protein powder is packaged in 50-pound bags that are loaded 25 to a pallet, so the most cost-effective way to order protein powder is by the pallet load (1,250 pounds).

Let's look at the materials requirements planning process using oats, which have a two-week lead time and must be ordered in hopper-truck quantities (multiples of 44,000 pounds). To determine when and how many pounds of oats should be ordered, we'll start with the weekly master production schedule for NRG-A and NRG-B bars, and then:

1. Convert the quantities from cases to 500-pound batches
2. Multiply the number of batches by the pounds-per-batch quantities (which are given in the BOM) to get the gross requirements for each raw material
3. Subtract the existing raw material inventory and purchase orders that have already been placed from the gross requirements, to determine the net requirements
4. Plan orders in multiples of the 44,000-pound lot size, allowing for the two-week lead time required for oats, to meet the net requirements in step 3

These steps are summarized in Figure 4-17. This view of the data is frequently called an **MRP record**, which is the standard way of viewing the MRP process on paper.

Oats Lead time = 2 weeks		Week 1	Week 2	Week 3	Week 4	Week 5
MPS	NRG-A	752	940	940	940	984
(cases)	NRG-B	322	403	403	403	422
MPS	NRG-A	108	135	135	135	142
(500 lb. batches)	NRG-B	46	58	58	58	61
Gross requirements (lb)		44,090	55,087	55,087	55,087	57,667
Scheduled receipts		44,000	44,000			
Planned receipts				88,000	44,000	44,000
On hand	11,650	11,560	473	33,386	22,299	8,632
Planned orders		88,000	44,000	44,000		

FIGURE 4-17   The MRP record for oats in NRG bars, weeks 1 through 5

The first two rows of the MRP record are the MPS that was the output from demand management (shown in Figure 4-14). These production quantities are in terms of shipping cases. The first step is to convert the MPS from shipping cases to 500-pound batches. Each shipping case weighs 72 pounds, so to convert shipping cases to 500-pound batches, multiply the number of shipping cases by 72 pounds per case, and then divide by 500 pounds per batch. Thus, producing 752 shipping cases of NRG-A bars in week 1 of the year will require 108 batches, as shown in Figure 4-17.

The next row in Figure 4-17 is gross requirements. The gross requirements figures are calculated by multiplying the MPS quantity (in production batches) by the pounds of oats needed for a batch of snack bars. FS uses 300 pounds of oats per batch of NRG-A bar and 250 pounds of oats per batch of NRG-B bar. This information is derived from the BOM (Figure 4-16). Therefore, for week 1, FS needs:

- NRG-A: 108.3 batches × 300 lb. per batch = 32,490 lb. oats
- NRG-B: 46.4 batches × 250 lb. per batch = 11,600 lb. oats

Total = 44,090 lb. oats

The next row in Figure 4-17 is the scheduled receipts. This row shows the expected arrival dates of orders of materials that have already been placed, meaning that the supplier has been given the purchase order and is in the process of fulfilling it. There is a two-week lead time for oats, so for oats to be available in week 1 and week 2 of the year, oats orders must be placed in the last two weeks of the previous year.

The next row, planned receipts, shows when planned orders will arrive. The planned receipts row is directly related to the planned orders row at the bottom of the record. A planned order is one that has not been placed with the supplier but will need to be placed to prevent Production from running out of materials. Because there is a two-week lead time for oat orders, the items in the planned orders will be available for production in two weeks, which is indicated by an entry in the planned receipts row. The arrows in Figure 4-17 show the relationship between planned orders and planned receipts. For example, the planned order for 88,000 pounds of oats in week 1 will be available for use in week 3, which is shown by the planned receipt of 88,000 pounds in week 3. The materials requirements planning calculation suggests that an order for 88,000 pounds of oats should be placed in week 1 so that it will arrive in week 3; there is only one order, but it shows up in two places on the MRP record.

The next row in Figure 4-17 is the on hand row. The first number in this row (11,650) is the inventory of oats on hand at the beginning of week 1. The number in the week 1 column (11,560) is a projection of the inventory that will be on hand at the end of week 1 (and therefore at the beginning of week 2)—accounting for the beginning inventory, gross requirements, and planned and scheduled receipts. In the case of the on-hand value for week 1, the initial inventory of 11,650 pounds, plus the 44,000 pounds of scheduled delivery, minus the 44,090 gross requirement, leaves 11,560 pounds of oats available at the start of week 2.

The last row is the planned orders row. This is the quantity that the MRP calculation recommends ordering, and it is the output from the MRP process that purchasing uses to determine what to order to produce the product, and when to order it.

Many times, a planner may intervene to tell the system to adjust the planned order. For example, notice that the on-hand quantity of oats in week 2 is only 473 pounds. This means that at the start of week 2, there will only be enough oats to mix one batch of dough. Since the production line produces 3,000 pounds of bars per hour, one batch of dough will keep the production line running for only 10 minutes. If the scheduled order does not arrive early enough on the first day of week 2, the production line could be shut down. When the purchase order scheduled to arrive in week 1 was ready to be placed (two weeks prior to the beginning of week 1), the planner should have evaluated that order, considering the low inventory level projected for the beginning of week 2. The planner might have decided to place an order for two hopper-truck loads of oats, instead of the planned order for one load. Or he could have ensured that the scheduled receipt shown in week 2 would actually be delivered at the end of week 1. Planning factors such as lead times are just estimates, so someone should evaluate the planned orders suggested by the materials requirements planning calculation before allowing the program to automatically turn them into purchase orders.

Notice once again the need for software to help with this kind of calculation. Of course, a human being can do these computations, but with many products and constituent materials, the calculations become very tedious and are prone to error. Even for a small company such as FS, doing the calculations with reasonable speed and accuracy requires software help. Notice also the information needed to do the MRP calculation: starting with a sales forecast, the software works down to the master production schedule and then to a schedule of needed raw materials.

Exercise 4.5

Develop an MRP record, similar to the one in Figure 4-17, for wheat germ for the five weeks of January. Wheat germ must be ordered in bulk-container quantities, so the planned orders must be in multiples of 2,000 pounds. Use a lead time of one week and an initial on-hand inventory of 1,184 pounds; assume that an order of 8,000 pounds is scheduled for receipt during week 1. Are there any weeks when you, as a planner, would consider placing an order above or below the minimum required? Why? Assume that there are no problems with storage capacity or shelf life.

Exercise 4.6

Fitter Snacker's purchasing policy has been to carry high levels of inventory to avoid stockouts. Why can inventory levels be lower with an integrated ERP system and MRP-based purchasing? If you had to calculate the financial advantage of this change, how would you do it?

Materials Requirements Planning in SAP ERP

The MRP list in SAP ERP looks very much like a Stock/Requirements List, which you already saw in Figure 4-12. The MRP list shows the results of the MRP calculations, while the Stock/Requirements List shows those results plus any changes that have occurred since the MRP list was generated (planned orders converted to purchase orders or production orders, material receipts, and so on). Because the materials requirements planning calculations are time-consuming to process for a company producing hundreds of products using thousands of parts, the MRP process is usually repeated every few days—or perhaps weekly. The Stock/Requirements List allows the users of the system to see what is happening (and will happen) with a material in real time. Compare the data shown in the MRP record in Figure 4-17 with the MRP list in Figure 4-18 and the Stock/Requirements List in Figure 4-19.

The MRP list in Figure 4-18 shows planned orders (PldOrd) and dependent requirements (DepR) in the second column. Dependent requirements represent the demand for snack bar dough from planned orders. Each batch of snack bar dough planned by the system requires 300 pounds of oats for NRG-A bars and 250 pounds of oats for NRG-B bars, which are referred to as dependent requirements because they are records (data) created by the production plans for snack bar dough. The MRP process creates planned

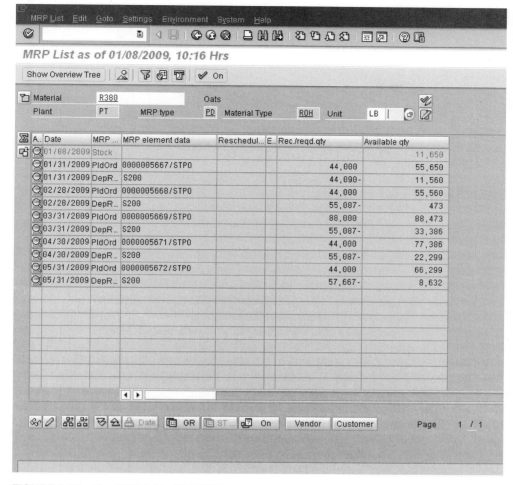

FIGURE 4-18 The MRP list in SAP ERP

orders to meet these dependent requirements. The planned orders are recommendations by the system to create orders (in this case, purchase orders) for oats.

The Stock/Requirements List shown in Figure 4-19 shows planned orders, but it also shows purchase requisitions (PurRqs) and purchase orders (POitem). When a planner decides that it is time for a planned order to become a purchase order, the planned order is converted to a purchase requisition, which is a request to Purchasing to create a purchase order. The planner can convert a planned order to a purchase order from the Stock/Requirements list by double-clicking the planned order line. This action calls up the window shown in Figure 4-20.

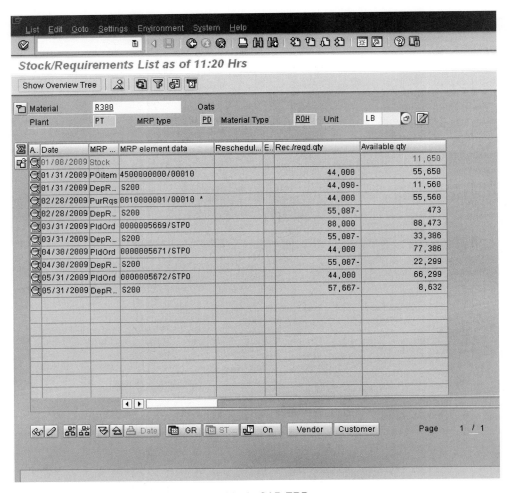

FIGURE 4-19   The Stock/Requirements List in SAP ERP

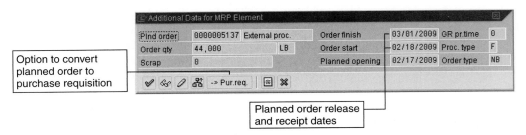

FIGURE 4-20   Conversion of planned order to purchase requisition

From this window, the planner can create a purchase requisition or review the planned order and make changes before creating the requisition. SAP ERP also provides the ability to mass-process planned orders, by converting groups of planned orders to purchase orders simultaneously. The materials requirements planning process can also be configured to automatically create purchase requisitions; for example, all planned orders created within one week of the MRP calculation could be automatically created as purchase requisitions.

Once a purchase requisition is created, an employee in the Purchasing department has to turn it into a purchase order. One of the important steps in creating a purchase order is choosing the best vendor to supply the material. An integrated information system such as SAP can facilitate this process. Figure 4-21 shows the Source Overview screen, which provides access to information that can help to select the vendor. From this screen, the Purchasing employee can view information about each vendor, simulate the price from each vendor (which might include quantity discounts and transportation costs), and also look at the vendor evaluation—the company's rating of the vendor. The SAP ERP system can be configured to rate vendors based on a number of performance criteria, including quality of goods provided and on-time delivery. The evaluation scores for each vendor are updated automatically as materials are received. The integrated information system allows Purchasing to make the best decision on a vendor based on relevant, up-to-date information.

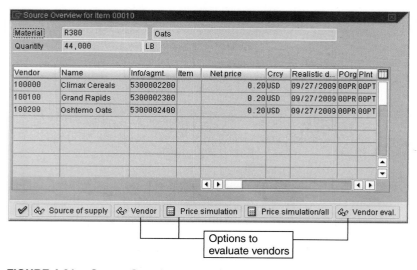

FIGURE 4-21   Source Overview screen for supplier selection

Once the Purchasing employee decides which vendor to use, the purchase order is transmitted to the vendor. The SAP ERP system can print out a paper order that can be mailed to the vendor. More likely, the system will be configured to either fax the order to the vendor, transmit it electronically through EDI (electronic data interchange), or send it over the Internet.

Detailed Scheduling

Finally, let's examine the last portion of the production process, detailed scheduling. The aggregate production plan for product groups developed in sales and operations planning is disaggregated to individual products in finer time increments through the demand management process. In detailed scheduling, a detailed plan of what is to be produced needs to be developed, considering machine capacity and available labor. Detailed scheduling is somewhat complex and tedious and will not be presented here in spreadsheet form, but we will discuss the important concepts and issues.

A key decision in detailed production scheduling is how long to make the production runs for each product. Longer production runs mean that fewer machine setups are required, reducing the production costs and increasing the effective capacity of the equipment. On the other hand, shorter production runs can be used to lower the inventory levels for finished products. Thus, the production run length requires a balance between setup costs and holding costs to minimize total costs to the company.

Because the capacity of the Fitter Snacker production mixers is much greater than that of the snack bar production line, scheduling mixer production is not an issue. Because the dough must be mixed before the snack bar production line can start, employees who run the mixers at FS should begin working a half-hour before the employees who run the production line. Therefore, four batches of dough can be mixed before the production line starts. With a bit of a head start and a detailed schedule for the production line, it is a simple matter for the personnel operating the mixers to keep ahead of the production line. Thus, the key step is to develop a detailed production schedule for the snack bar production line.

The manufacturing process that Fitter Snacker uses is known as repetitive manufacturing. **Repetitive manufacturing** environments usually involve production lines that are switched from one product to another similar product. Most packaged consumer goods are produced in repetitive manufacturing environments. In repetitive manufacturing, production lines are scheduled for a period of time, rather than for a specific number of items, although it is possible to estimate the number of items that will be produced over a period of time. For Fitter Snacker, the production schedule for a week might be to produce NRG-A bars from Monday morning until the end of Wednesday, then change over to NRG-B bars for all of Thursday until Friday at noon, when the production line will be switched back to NRG-A bars. Given this schedule, it is possible to estimate the number of bars that will be produced. Figure 4-22 shows the repetitive manufacturing planning screen in the SAP ERP system. This screen allows the planner to view capacity, production schedule length, and quantity produced in one screen. NRG-A bars are being produced for 3½ days, and NRG-B bars are being produced for 4½ days. The top line shows available capacity of 24 hours a day from September 15 through September 28. The bottom screen shows that a total of 1,095 cases of NRG-A snack bars will be produced from September 15 through September 20, and then production will change over to NRG-B bars on September 20. The NRG-B bars will be produced through September 24. Notice that Fitter Snacker has no production scheduled over the weekend.

In some companies, responsibility for inventory costs belongs to a Materials Management group, and capacity utilization performance is the responsibility of a Production group. The Materials Management group wants short production runs to keep inventory

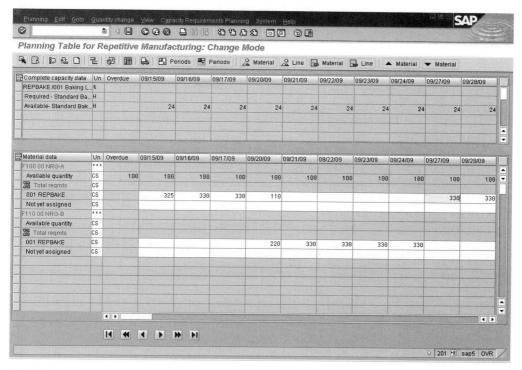

FIGURE 4-22   Repetitive manufacturing planning table in SAP ERP

levels down, while the Production group wants long production runs to keep capacity utilization high. In these circumstances, the decision regarding production run length can become a source of organizational bickering, instead of a decision that minimizes total costs for the benefit of the company.

This conflict points out an advantage of production planning in an ERP system. Because the goal of the company is to maximize profit, the duration of production runs should be decided by evaluating the cost of equipment setup and holding inventory. An integrated information system simplifies this analysis by automatically collecting accounting information that allows managers to better evaluate schedule trade-offs in terms of costs to the company.

Providing Production Data to Accounting

In Chapter 1 you learned that functional areas must share data for a company to be successful. Accounting needs to know what Manufacturing has produced and what resources were used in producing those products, to determine which products, if any, are producing a profit—and then provide information for management to determine how to increase profits. In the manufacturing plant, ERP packages do not directly connect with production machines. For example, in Fitter Snacker's case, SAP ERP could not directly read the number of bars that came off the packing segment of the snack bar line (Figure 4-1). The data must be gathered in some way and then entered into SAP ERP for inventory accounting purposes.

Data can be entered into SAP ERP through a PC on the shop floor, scanned by a barcode reader, or entered using a wireless PDA. SAP ERP is an open-architecture system, meaning that it can work with automated data-collection tools marketed by third-party hardware and software companies. Radio frequency identification (RFID) technology has simplified this process. RFID technology is discussed in more detail in Chapter 8.

In an integrated ERP system, the accounting impact of a material transaction can be recorded automatically. For example, when a shipment of oats arrives at the Fitter Snacker plant, someone in the Receiving department must verify the material and the quantity and quality of the shipment before it is accepted. Once FS accepts the shipment, Receiving must notify the SAP ERP system of the arrival and acceptance of the material. This communication is done by completing a goods receipt transaction, which is shown in Figure 4-23.

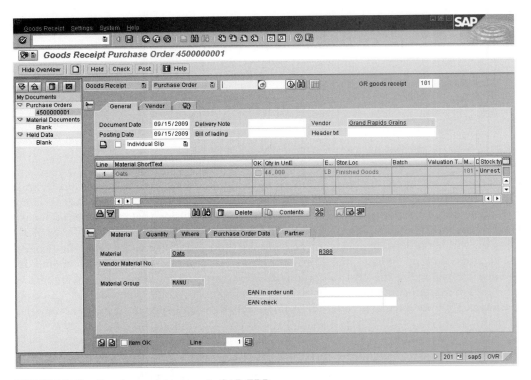

FIGURE 4-23 Goods receipt screen in SAP ERP

The Receiving department must match the goods receipt with the purchase order that initiated it, to make sure that the exact materials ordered have been received, so that Accounting can pay the vendor. It is possible for the quantity of material entered in the goods receipt to differ from the quantity specified on the purchase order. Depending on the configuration settings, the SAP ERP system might block entry of the receipt if the discrepancy is too large. If the discrepancy is small, then the receipt may be allowed, with the difference posted to the correct variance account, which allows the transaction to be processed but maintains a record for management to review, to see if there is a consistent problem with a vendor "shorting" an order (consistently shipping less than was ordered).

When the receipt is successfully recorded, the SAP ERP system immediately records the increase in inventory levels for the material. On the Accounting side of the system, this causes the value of the inventory shown in the general ledger account to automatically increase as well. This is an important feature of an integrated information system: the goods receipt is recorded once, but the information is immediately available to both Manufacturing and Accounting—and the information is consistent. An integrated information system also has the ability to adjust for changes in material costs. If the cost of the material changes frequently, the system can be configured to reevaluate the value of all the inventory of the material that the company has. For example, suppose Fitter Snacker has 10,000 pounds of cinnamon that it bought at $3 per pound. The company would value its inventory of cinnamon at $30,000. Then suppose the price of cinnamon has risen to $4 per pound, and Fitter Snacker has just purchased 1,000 pounds of cinnamon at the higher price. What is the value of the 11,000 pounds of cinnamon that Fitter Snacker now owns? The SAP ERP system can be configured to use a moving average formula to reevaluate the inventory based on the current market prices. Depending on the exact nature of the formula, the cinnamon would be valued at somewhere between the $3 per pound previously assumed and the $4 per pound that was just paid. The system can perform this calculation automatically each time material is received.

Using an ERP package to record data does not necessarily make the shop-floor accounting data more accurate. The ERP system allows employees to enter data in real time. Furthermore, capturing data for manufacturing and inventory purposes on the shop floor means that it is captured at the same time for accounting and inventory management purposes—eliminating any need to reconcile Accounting and Manufacturing records. But the system requires employees to follow the process. If employees can take material out of inventory without recording the transaction, then the real-time information in the ERP system is worthless. Technologies such as bar-code scanners and RFID tags can help in this process, but accurate data require both a capable information system and properly trained and motivated employees.

Exercise 4.7

Briefly describe how the implementation of SAP ERP might change the relationship between Production and Warehousing at Fitter Snacker.

ERP AND SUPPLIERS

As mentioned in the introduction to this chapter, Fitter Snacker is part of a supply chain that starts with farmers growing oats and wheat and ends with a customer buying an NRG bar from a retail store. Previously, companies used competitive bidding to achieve low prices from suppliers, which frequently led to an adversarial relationship between suppliers and their customers. In recent years, companies have realized that they are part of a supply chain, and if the supply chain is more efficient, all participants in the chain can benefit. Collaboration can frequently achieve more than competition, and ERP systems can play a key role in collaborative planning.

Working with suppliers in a collaborative fashion requires trust among all parties. A company opens its records to its suppliers, and suppliers can read the company's data because of common data formats. Working this way with suppliers cuts down on paperwork and response times. Reductions in paperwork, savings in time, and other efficiency improvements translate into cost savings for the company and the suppliers. ERP lets companies and suppliers share information (sales, inventory, production plans, and so on) in real time throughout the supply chain. This allows all parties (suppliers, manufacturers, and customers) to eliminate from the supply chain costs that don't add value to the product (such as inventory, overtime, changeovers, and spoilage), while simultaneously improving customer service.

The Traditional Supply Chain

The term **supply chain** describes all of the activities that occur between the growing or mining of raw materials and the appearance of finished products on the store shelf. In a traditional supply chain, information is passed through the supply chain reactively as participants increase their product orders—as illustrated in Figure 4-24. For example, a retailer sees an increase in the sales of FS's bars and orders a larger quantity of bars from the wholesaler. If a number of retailers increase their orders, the wholesaler will increase its orders from Fitter Snacker. When FS gets larger orders from wholesalers, it must increase production to meet the increased demand. To increase production, FS will order more raw materials from suppliers.

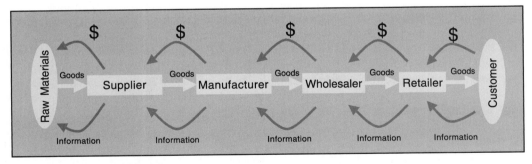

FIGURE 4-24   Supply chain management (SCM) from raw materials to consumer

Because of the time lags inherent in a traditional supply chain, it might take weeks—or even months—for information about FS's increased need for raw materials to reach FS's suppliers. Raw material suppliers may require time to increase *their* production to meet FS's larger orders, resulting in temporary shortages for the supplier.

By contrast, if the participants in the supply chain are part of an integrated process, information about the increased customer demand can be passed quickly through the supply chain, so each link in the chain can react quickly to the change.

EDI and ERP

The development of supply chain strategies does not necessarily require an ERP system. Before ERP systems were available, companies could be linked with customers and suppliers through electronic data interchange (EDI) systems. Recall from Chapter 2 that EDI is the computer-to-computer exchange of standard business documents (such as purchase orders) between two companies. A well-developed ERP system, however, can facilitate SCM because the needed production planning and purchasing systems are already in place. In addition, the integration of accounting data in the ERP system (described in the next chapter) allows management to evaluate changes in the market and make decisions about how those changes should affect the production plan. With an ERP system, sharing production plans along the supply chain can occur in real time. Using the Internet can make this communication even faster and cheaper than using private EDI networks.

ANOTHER LOOK

Supply Chain Management: It's All About Relationships

Procurement, the ordering of raw materials and supplies, has the potential to save companies millions of dollars. According to FinListics, an Atlanta consulting company, if an average Standard & Poor's-listed company with $5 billion in revenue could reduce cost through better procurement by 5 percent, it could increase profits by $20 million. The network technology; namely, the Internet, is in place. When a customer places an order on the Internet, it starts a chain of reaction from all suppliers and manufacturers to fill that order. With the network technology in place, companies need only manage the relationships that go into these savings. Companies must open themselves up and share their sales forecasts, and manufacturing and inventory management systems. Traditionally, companies' suppliers and manufacturers have had adversarial relationships rather than open ones.

One company, Cisco, has taken a friendlier approach with its suppliers, and has modeled itself after the Japanese automakers – enabling their suppliers rather than challenging them. Cisco's philosophy is that if it can help suppliers be more efficient, it too can be more efficient. Cisco, for example, operates an extranet to include suppliers on future product plans so suppliers can start to gear up for future orders. Cisco, which makes networking equipment, handles over 90 percent of orders over the Internet. Less than half of Cisco employees even touch an order. Incoming orders trigger alerts to suppliers that make components for that specific order. Throughout order assembly, Cisco's manufacturing system is checking to make sure all assembly processes are set up perfectly. The supply chain basically runs itself. According to AMR Research, Cisco's revenue was $713,000 per employee compared with the average $192,000 per employee of a Fortune 500 company in 1999. Cisco says its administrative overhead has gone from $100 per order to between $5 and $6 per order.

With this sharing of information, ERP systems are more important than ever, according to Tim Lambeth, the *Vanity Fair* vice president of global finance. "[ERP is] more critical than ever. If you are going to begin to collaborate with your suppliers, you will have

continued

to have real-time information available to them. If Wal-Mart wants to come into my system to place or track orders, it expects my system to tell it precisely what I can do and when I can do it." Lambeth states that companies who have not used the Web for their supply chain management will soon be forced to by their customers.

Supply chain management comes down to relationships. Communication between trading partners is critical. For example, Staples, the office supply chain, wanted to know why Hewlett-Packard required a 21-day lead time for orders. HP had believed that Staples required the 21-day lead time, when in fact neither company required it – the parameter was some leftover decision from the days when purchasing agents cut deals. Now the two companies have a seven-day lead time for orders. Staples can hold less safety stock, has fewer stock-outs, and HP has reduced its order-to-cash cycle time.

Staples shares a lot of information with its suppliers. In 2005, Staples set up a secure Web site where its suppliers can log on and view Staples supply chain metrics. The metrics act as a report card for each company, allowing for further communication with suppliers and thus improvement with the supply chain.

Question:

1. How can Fitter Snacker improve its relationships with its suppliers of raw materials?

The Measures of Success

Performance measurements (sometimes referred to as **metrics**) have been developed to show the effects of better supply chain management. One measure is called the **cash-to-cash cycle time**. This term refers to the time between paying for raw materials and collecting cash from the customer. In one study, the cash-to-cash cycle time for companies with efficient SCM processes was a month, whereas it was 100 days for companies without SCM.

Another measure is total SCM costs. These costs include the cost of buying and handling inventory, processing orders, and information systems support. In one study, companies with efficient SCM processes incurred costs equal to 5 percent of sales. By contrast, companies without SCM incurred costs of up to 12 percent of sales.

Other metrics have been developed to measure what is happening between a company and its suppliers. For example, Staples, the office-supply company, measures three facets of the relationship. **Initial fill rate** is the percentage of the order that the supplier provided in the first shipment. Another metric is **initial order lead time**, which is the time needed for the supplier to fill the order. Finally, Staples measures **on-time performance**. If the supplier agreed to requested delivery dates, this measurement tracks how often the supplier actually met those dates.

Improvement in metrics like these leads to improvement in overall supply chain cost measurements.

Exercise 4.8

Assume a manufacturer of residential lawn and garden equipment is considering investing in hardware and software that will improve linkages with suppliers. Management expects to save 5 percent of sales by tightening up the supply chain in the first year, 3 percent in the second year, and 1 percent in the third year. The company's annual sales are $1 billion. The company's chief financial officer insists that the investment must pay for itself in cost savings in three years. To meet this requirement, how much should the chief information officer be allowed to spend on improving the supply chain? Explain your answer.

ANOTHER LOOK

SCM and Its Critical Success Factors

For a supply chain management project to be successful, a company must achieve certain key factors. The Advisory Council at *Information Week* has put together a list of those key factors:

- *Business-driven strategy:* The information system for managing the supply chain must focus on the customer, allowing the customer greater efficiencies in the process. The customer should find that ordering material from the vendor is now more efficient with the new system. The system is based on a complex business process model. When a vendor fulfills a customer's order, many different business processes are at work to ensure quality products, timely deliveries, and good prices.
- *Management commitment:* As with any IT project, top management must be committed to support the project; otherwise, the project is doomed to fail.
- *Supplier and partner expectation management:* An SCM implementation requires change from all parties. A company implementing an SCM project must be prepared to work closely with both vendors and customers and manage that change from all angles.
- *Internal expectation management:* Change management is especially important within a company. Personnel involved in a new system must be informed and trained to manage any anxiety they might experience in their job changes.
- *Learning period acceptance:* Companies implementing SCM must accept that it takes a long time to fully realize the benefits of the system.

SCM has the further benefit of driving business value. Simply put, SCM will help your company's bottom line by making it more efficient for customers to buy your product and for you to buy your suppliers' products. In this way, your entire company will benefit and become more efficient.

Implementing an SCM system without paying attention to the critical factors listed above will result in poor performance. In fact, a large percentage of companies that implement SCM are not satisfied. Forty-five percent of respondents to a Booz Allen Hamilton supply chain management survey report that their IT solutions are failing to meet expectations. SCM consultants often find that companies have unrealistic expectations of the technology. However, companies dealing in the service parts business (for example, Caterpillar Logistics and Ford Motor Company) are finding that SCM allows them to

continued

decrease inventory by between 10 and 30 percent, according to AMR Research Inc. Companies must decide how their supply chains should be designed and run, and then use technology to implement those goals.

Questions:

1. Assume you were recently hired in an IT role at a large manufacturing company. Your first job is to convince the manufacturing manager that the implementation of an SCM system is feasible. Write a memo to that manager, highlighting advice on how to ensure the smooth implementation of such a system.

2. Outline the potential pitfalls that your company may encounter when implementing SCM.

ANOTHER LOOK

Nike's Global Planning Initiative

Nike has spent over eight years and $500 million on its supply chain project, which was originally planned to take six years and cost $400 million. In addition to increased implementation costs, the project also cost Nike more than $100 million in lost sales, reduced its stock price by 20 percent, triggered a flurry of class-action lawsuits, and caused chairman, president, and CEO, Phil Knight, to publicly lament, "This is what you get for $400 million, huh?"

Nike's original business model, developed in the 1970s, was based on centralization. In the 1970s, U.S. sneaker manufacturers were just beginning to produce shoes in the Far East. While its competitors were suffering from unpredictable delivery and high levels of inflation, Nike dominated the market by guaranteeing delivery and locking in an inflation-proof discount from its Asian suppliers. Nike, in return, committed to placing orders six months in advance. Both retailers and Nike gained certainty with this arrangement.

But as Nike grew and times changed, Nike's central control became more decentralized. By 1998, Nike had 27 different order management systems worldwide, all poorly linked to the Beaverton, Oregon, facility that was supposed to coordinate all planning. To regain control over production planning, Nike selected SAP ERP to be its information system in conjunction with i2's supply, demand, and collaboration planning software.

The implementation project began on a troubled path. Nike decided to deploy the i2 software with its legacy systems prior to implementing SAP ERP. The i2 software required significant customization to make it compatible with Nike's legacy systems, and the software didn't always work. Sometimes it took as long as a minute for a single transaction

continued

to be processed, and the tens of millions of product numbers Nike used were too much for the i2 software, causing the system to crash frequently. The i2 system ignored some production orders, duplicated others, and deleted ordering data six to eight weeks after it was entered, making it impossible for planners to recall what they had asked each factory to produce. Finally, the software instructed Nike to make a certain number of different types of sneakers—instructions that often conflicted with the real demand from the customer. Soon the Asian factories were producing products that had no actual match with customer demand.

Another big problem with Nike's implementation was a lack of appreciation for the importance of the project. At $10 million, the price tag of the i2 system represented a small percentage of the total $400 million supply chain project, leading to a feeling that it would be an easier part of the project. According to Nike CIO Gordon Steele, "This felt like something we could do a little easier since it wasn't changing everything else [in the business]. But it turned out it was very complicated."

Technology was not the only source of problems with the implementation. Asked whether more user training would have helped, Steele replied, "You can never train enough." Nike now makes training an important part of its business. For example, U.S. customer service representatives are locked out of the SAP ERP system until they complete the full training course, which consists of 140 to 180 hours of instruction from highly trained fellow Nike employees.

Question:

1. Suppose you were responsible for implementing new supply chain planning software for Fitter Snacker. How would you plan the project to make sure that the system would deliver reliable plans? Whom would you put on the implementation team? How would you test the system? What training would you provide for employees, and when?

Chapter Summary

- An ERP system can improve the efficiency of production and purchasing processes. Efficiency begins with Marketing sharing a sales forecast. A production plan is created based on that forecast and shared with Purchasing so raw materials can be ordered properly.

- Companies can do production planning without an ERP system, but an ERP system that contains materials requirements planning allows Production to be linked to Purchasing and Accounting. This data sharing increases a company's overall efficiency.

- Companies are building on their ERP systems and integrated systems philosophy to practice supply chain management (SCM), a strategy by which a company looks at itself as part of a larger process that includes customers and suppliers. Using information more efficiently along the entire chain can result in significant cost savings. Because of the complexity of the global supply chain, developing a planning system that effectively coordinates information technology and people is a considerable challenge.

Key Terms

Bill of material (BOM)	Metrics
Capacity	MRP record
Cash-to-cash cycle time	On-time performance
Initial fill rate	Repetitive manufacturing
Initial order lead time	Rough-cut planning
Lead time	Standard costs
Lot sizing	Supply chain
Master production schedule (MPS)	

Exercises

1. In which industries is Supply Chain Management important? In which industries is it not? Why, or why not?

2. Review the information in this chapter regarding Fitter Snacker's production problems, and create a flowchart to document that flawed process. Use the flowcharting tools in Microsoft Excel to produce a professional-looking diagram. Supplement your flowchart with a paragraph that explains how ERP systems will help remove flaws from the FS production process.

3. Using an Internet search engine or the library, research a manufacturing company that has successfully implemented SCM. Did the company have an ERP system prior to the SCM implementation? Can the company quantify the benefits of having SCM? How does the company measure success?

4. Summarize the sources and destinations of production and purchasing information within a company that has MRP and ERP. Refer to Figure 4-2 as a guide. Data could be summarized in a tabular format using the table feature in a word processor. Use three columns: Data, Source, and Provided to.

5. Interview a professor at your university who teaches in the operations research area. Ask the professor about his or her industry contacts and how those contacts use ERP, MRP, and SCM. Get the name of a production manager in a manufacturing firm, and then try to make an appointment with that manager. If possible, interview that production manager and ask about the flow and channels of information to and from the Production department.

6. Compare Customer Relationship Management (CRM) and SCM. How are they similar? How are they different? Where do they interact? In answering, consider the kinds of technologies used in each.

For Further Study and Research

Bacheldor, Beth. "Supply Chain Management Still A Work In Progress." *Information Week,* May 19, 2003. http://www.informationweek.com/showArticle.jhtml; jsessionid=NKXLRK21WJW24QSNDLRSKH0CJUNN2JVN?articleID=10100101.

Beg, Humayun, Stephen Rood, and Bill Spernow. "Smart Advice: Top-Down Strategy." *Information Week,* February 23, 2004. http://www.informationweek.com/showArticle.jhtml; jsessionid=NKXLRK21WJW24QSNDLRSKH0CJUNN2JVN?articleID=17701385.

Koch, Christopher. "Nike Rebounds: How (and Why) Nike Recovered from Its Supply Chain Disaster." *CIO,* June 15, 2004. http://www.cio.com/article/32334/.

———. "The Big Payoff." *CIO,* June 12, 2007. http://www.cio.com/article/118901/The_Big_Payoff.

———. "The Big Payoff." *CIO,* October 1, 2000.

Sullivan, Laurie, "SAP Launches Latest Supply Chain Management Suite." *Information Week,* March 9, 2006. http://www.informationweek.com/showArticle.jhtml; jsessionid=EVB2F0BHSAHFQQSNDLRSKH0CJUNN2JVN? articleID=181502558&queryText=scm+benefits.

CHAPTER 5

ACCOUNTING IN ERP SYSTEMS

LEARNING OBJECTIVES

After completing this chapter, you will be able to:

- Describe the differences between financial and managerial accounting.
- Identify and describe problems associated with accounting and financial reporting in unintegrated information systems.
- Describe how ERP systems can help solve accounting and financial reporting problems in an unintegrated system.
- Describe how the Enron scandal and the Sarbanes-Oxley Act have affected accounting information systems.
- Explain accounting and management-reporting benefits that accrue from having an ERP system.

INTRODUCTION

In previous chapters, you learned about functional area activities, both generally and specifically: In Chapter 1, you read an overview of functional area activities; in Chapter 3, you learned about Marketing and Sales activities; and in Chapter 4, you learned about Supply Chain Management. In this chapter, you will learn about the activities in another functional area, Accounting. You will see how Accounting is tightly integrated with all of the other functional areas, as well as how Accounting activities are necessary for decision making.

ACCOUNTING ACTIVITIES

Accounting activities can generally be classified as either financial accounting or managerial accounting. An additional area of accounting, tax accounting, is beyond the scope of this text. Because tax accounting is chiefly the external reporting of a business's activities to the Internal Revenue Service, data gathered for financial accounting forms the basis for tax accounting.

Financial accounting consists of documenting all transactions of a company that have an impact on the financial state of the firm, and then using these documented transactions to create reports for external parties and agencies. These reports, or financial statements, must follow the prescribed rules and guidelines of various agencies, such as the Financial Accounting Standards Board (FASB), the Securities and Exchange Commission (SEC), and the Internal Revenue Service (IRS).

Common financial statements include balance sheets and income statements. The **balance sheet** is a statement that shows account balances such as cash held, amounts owed to the company by customers, the cost of raw materials and finished-goods inventory, long-term assets such as buildings, amounts owed to vendors, amounts owed to banks and other creditors, and amounts that the owners have invested in the company. A balance sheet is a good overview of a company's financial health at a point in time, a key consideration for a company's creditors and owners. Figure 5-1 shows a balance sheet for Fitter Snacker.

Fitter Snacker Balance Sheet At December 31, 2009 (in thousands of dollars)		
Assets		
Cash		$5,003
Accounts receivable		$4,715
Inventories		$9,025
Plant and equipment		$6,231
Land		$1,142
Total assets		$26,116
Liabilities		
Accounts payable	$6,400	
Notes payable	$10,000	
Total liabilities		$16,400
Stockholders' Equity		
Contributed capital	$2,000	
Retained earnings	$7,716	
Total stockholders' equity		$9,716
Total liabilities and stockholders' equity		$26,116

FIGURE 5-1   Fitter Snacker sample balance sheet

The **income statement**, or **profit and loss (P&L) statement,** shows the company's sales, cost of sales, and the profit or loss for a period of time (typically a quarter or year). Profitability is important to creditors and owners. It is also important information for managers in charge of day-to-day operations. A manager sees profits as indicators of success and losses as indicators of problems to be solved. You can get an idea of the contents of an income statement by looking at Figure 5-2, which shows an income statement for Fitter Snacker.

Fitter Snacker Income Statement For the Year Ended December 31, 2009 (in thousands of dollars)		
Revenues		
Sales revenue	$36,002	
Total revenues		$36,002
Expenses		
Cost of goods sold expense	$25,691	
Selling, general, and administrative expense	$4,251	
Research and development expense	$962	
Interest expense	$521	
Total expenses		$31,425
Pretax income		$4,577
Income tax expense		$1,144
Net income		$3,433

FIGURE 5-2   Fitter Snacker sample income statement

Companies prepare financial statements quarterly, and sometimes more frequently. To prepare these statements, companies must "close their books," which means that temporary or nominal accounts (such as revenue, expense, gain, and loss) have their balances transferred or closed to the retained earnings account. The closed nominal accounts will have a zero balance from which to start accumulating revenues and expenses in the next reporting period. Closing entries are made to transfer balances and to establish a zero balance. To do this, employees must check the accounts to see that they are accurate and up to date. If a company's information systems routinely generate accurate and timely data, closing the books can go smoothly. If they do not, "adjusting" entries must be made, and, in that case, closing the books can be a very time-consuming chore with inaccurate results.

One advantage of an integrated information system is that it simplifies the process of closing the books and preparing financial statements. There is no need to assemble data from different systems because all of the required data are contained in a centralized system. Figure 5-3 shows how Fitter Snacker's balance sheet and P&L statement would look in the SAP ERP system. In an ERP system, the balance sheet and P&L statement are database reports. They can be quickly generated at any time, and because the data to prepare the reports are read from the database tables, these reports are always up-to-date. Another feature of the ERP balance sheet and P&L statement is the ability to quickly display data at different levels of detail, as shown in Figure 5-4. In addition, the system allows the user to create financial statement variants, which are financial statements in other formats, prepared to suit the needs of different users.

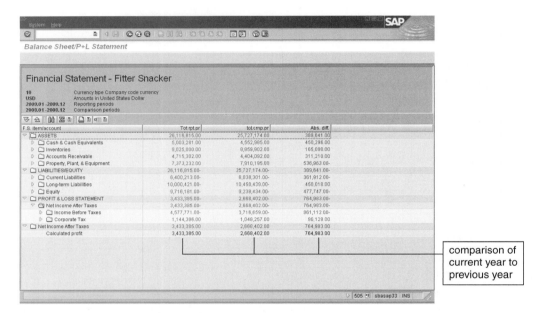

FIGURE 5-3   Balance sheet and income statement for Fitter Snacker in SAP ERP system

Managerial accounting deals with determining the costs and profitability of the company's activities. While the information in a company's balance sheet and income statement shows whether a firm is making an overall profit, the goal of managerial accounting is to provide managers with detailed information that allows them to determine the profitability of a particular product, sales region, or marketing campaign. Managerial accounting provides information that managers use to control a company's day-to-day activities and to develop long-term plans for operations, marketing, personnel needs, repayment of debt, and other management issues. Because managerial accounting provides reports and analyses for internal use, companies have great flexibility in how they configure their managerial accounting systems.

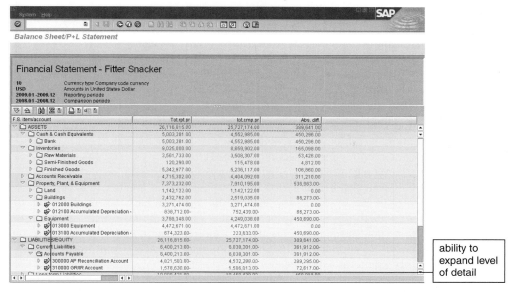

FIGURE 5-4 Balance sheet and income statement in SAP ERP, showing ability to control level of detail

Using ERP for Accounting Information

Recall from Chapter 1 that in the past, companies have had separate functional information systems: a marketing information system, a manufacturing information system, and so on, each with its own way of gathering data and its own file system for recording data. Companies built these unintegrated systems primarily to handle the needs of the individual functional areas, and secondarily to provide data to Accounting. With unintegrated systems, the functional areas shared their data with Accounting, so Accounting could "keep the books," that is, maintain records of all financial transactions. Data sharing, however, usually did not occur in real time, so Accounting's data were often out of date. Further, because the shared data might not be the only information that Accounting needed to prepare reports for management, accountants, and functional area clerks had to do significant research. Since the 1960s, legions of accountants, analysts, and programmers have tried to make unintegrated systems work, but this approach has not worked very well, as prior chapters illustrate.

An ERP system, with its centralized database, avoids these problems. For example, suppose finished goods are transferred from the assembly line to the warehouse. An employee in the warehouse can easily record the transaction, using a terminal or a bar-code scanner. In SAP ERP, the Materials Management module would see the transfer event as an increase in finished-goods inventory available for shipment; the Accounting module would see the event as an increase in the monetary value of finished goods. With ERP, everyone uses the same database to record operating data. This database is then used to generate management reports, make financial statements, and create budgets.

In traditional accounting, a company's accounts are kept in a record called the **general ledger**. In the SAP ERP system, input to the general ledger occurs simultaneously with the business transaction in the specific module. Many SAP ERP modules cause transaction data to be entered into the general ledger, including:

- Sales and Distribution (SD), which lets the user record sales and creates an **accounts receivable** entry (a general ledger document that indicates a customer owes money for the goods received).
- Materials Management (MM), which controls purchasing and recording of inventory changes. The creation of a purchase order creates an accounts payable entry in the general ledger, noting that the company has an obligation to pay for goods that it will receive. Whenever material moves into or out of inventory (purchased materials arrive from the vendor, materials are taken out of inventory to support production, or finished goods go from production to inventory), general ledger accounts are affected.
- Financial Accounting (FI), which manages the accounts receivable and accounts payable items created in the SD and MM modules, respectively. The FI module is also where the general ledger accounts are closed at the end of a fiscal period (quarter or year) and financial statements are generated.
- Controlling (CO), which tracks the costs associated with producing products. To make a profit, it is critical for the company to have an accurate picture of its product costs, allowing it to make correct decisions about product pricing and promotions, as well as capital investments.
- Human Resources (HR), which manages the recruiting, hiring, compensation, termination, and severance of employees. The HR module also manages benefits and generates the payroll.
- Asset Management (AM), which manages fixed-asset purchases (plant and machinery) and the related depreciation.

OPERATIONAL DECISION-MAKING PROBLEM: CREDIT MANAGEMENT

Out-of-date or inaccurate accounting data that result from an unintegrated information system can cause problems when a company is making operational decisions. This problem was illustrated in Chapter 3 by Fitter Snacker's challenges in making credit decisions. In this section, FS's credit-granting problems will be discussed in more detail. First, we'll look at industrial credit granting in general, and then at FS's credit-check problem.

Industrial Credit Management

Companies routinely sell to customers on credit. Good financial management requires that only so much credit be extended to a customer, however. At some point, the customer must pay off some of the debt to justify the faith the seller has shown (and so that the seller can turn the accounts receivables into cash). Credit management requires a good balance between granting sufficient credit to support sales and making sure that the company does not lose too much money by granting credit to customers who end up defaulting on their credit obligations.

In practice, sellers manage this relationship by setting a limit on how much money a customer can owe at any one time, and then monitoring that limit as orders come in and payments are received. For example, the seller might tell a buyer that her credit limit is $10,000, which means that the most she can owe the seller is $10,000. If the buyer reaches that amount, the seller will accept no further sales orders until she pays off some of her debt. When making a sale on credit, the seller makes an entry on the books to increase his accounts receivable and his sales. Thus, when the buyer's accounts receivable balance on the seller's books reaches $10,000, the buyer must make some payment.

Continuing the example, assume that the buyer calls the seller to order $3,000's worth of goods, and her credit limit is $10,000. If the seller's records show that the accounts receivable balance for the buyer is already $8,000, then the seller should not accept the $3,000 order, because it would bring the accounts receivable balance to $11,000, which exceeds the buyer's credit limit. Instead of refusing the order, the seller's sales representative might suggest that the buyer reduce the size of the order, or ask her to send in a payment before processing the order, thus reducing the buyer's debt. Clearly, to make this system work, a sales representative needs to be able to review an up-to-date accounts receivable balance when an order comes in.

If Marketing and Accounting have unintegrated information systems, full cooperation between the two functional areas will not be easy to achieve. Marketing knows the current order's value, but Accounting keeps the accounts receivable records. If Accounting keeps the books up-to-date and can provide the current accounts receivable balance to Marketing *when needed*, then credit limits can be properly managed. Marketing can compare the customer's credit limit to an accurate balance-owed amount (plus the order's value) to make a decision. However, in an unintegrated system, Accounting may not immediately record sales and/or payment receipts as they occur. In that case, accounts receivable balances will not be current. Furthermore, the sales clerk may be working from an out-of-date credit-balance printout. If the printout balances do not reflect recent payments, a customer may be improperly denied credit. The customer would probably challenge the denial, which would trigger a request for updated information in Accounting. The delay entailed in that research could reduce customer satisfaction, and performing the research would consume valuable employee time.

These problems should not arise with an integrated information system. When a sale is made, the system immediately increases accounts receivable. When the company receives and records a payment, accounts receivable is immediately decreased. Because the underlying database is available to Marketing *and* Accounting, sales representatives can also see customer records immediately. Thus, sales representatives do not need to make a request to Accounting for the customer's accounts receivable balance.

With that background, we can now consider how Fitter Snacker handles credit management.

Fitter Snacker's Credit Management Procedures

As described in Chapter 3, an FS sales clerk refers to a weekly printout of a customer's current balance and credit limit to see if credit should be granted. Assuming that the customer's order would not present credit-limit problems, the sales clerk enters the sale in the sales order entry system, which is a stand-alone computer program. Sales data are transferred to Accounting by disk three times a week. An accounting clerk can use the sales input to prepare a customer invoice.

Accounting must make any adjustments for partial shipments before preparing the invoice. The accuracy of the adjustment process depends on whether the warehouse transmits order changes in a timely fashion. Accounting also makes the standard revenue-recognition accounting entry: a debit to accounts receivable and a credit to sales for the amount billed.

Accounting clerks also process customer payments. Clerks receive and manually handle checks. They enter data in the accounting program, increasing cash and decreasing accounts receivable. These data are later used to make entries to individual customer accounts, reducing the amount that customers owe to FS. If time permits, accounts are posted on the day payment is received and the bank deposit is made; otherwise, the entries are done as soon as possible the next day. Thus, there can be some delay between the time FS receives a check from a customer and the actual reduction of the customer's accounts receivable balance.

Now let's look at how SAP ERP could help FS's credit management.

Credit Management in SAP ERP

Suppose FS is using SAP ERP as its ERP system. This system would allow FS to set a credit limit for each customer. A company can configure any number of credit-check options in the SAP ERP system. Figure 5-5 shows a dynamic credit check with Reaction C selected. Reaction C means that if the order being saved will cause the customer to exceed its credit limit, the system will issue a warning indicating the amount by which the order exceeds the credit limit. Because the system is issuing a warning, the order can be saved, but will be blocked from further processing until the credit problem is cleared. Frequently, companies do not configure the system to provide warnings to sales order clerks because they are not equipped to correct the problem and because the credit problem is an issue between the selling firm's Accounts Receivable department and the customer's Accounts Payable department.

Figure 5-5 also shows that the credit check is Dynamic and has a two-month horizon. This means that only the next two months of sales orders will be used in calculating the credit check. Customers may place orders for a long-range schedule, but only those that will be shipped in the near term are usually considered in the credit check.

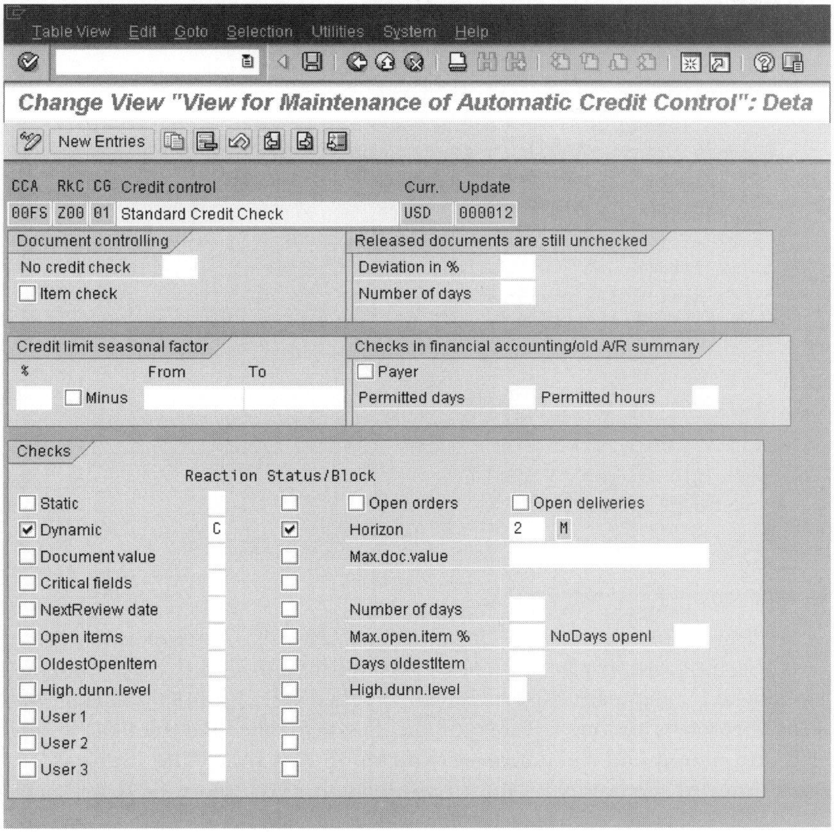

FIGURE 5-5 Credit management configuration

Figure 5-6 shows the credit-checking process in Figure 5-5 applied to a specific customer, Health Express. Health Express has a credit limit of $1,000 and currently has used $590 of this limit. If Health Express places an order for snack bars that totals more than $410, then the order will be blocked.

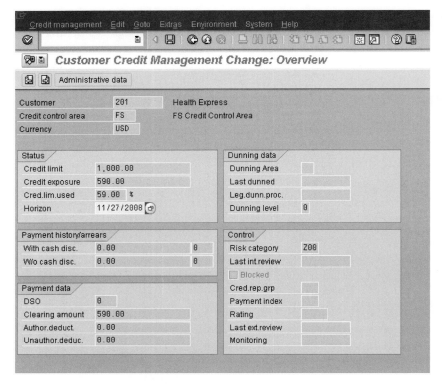

FIGURE 5-6   Credit management for Health Express

Figure 5-7 shows the transaction where blocked sales orders are listed. Most companies have an employee who is responsible for reviewing blocked sales orders (perhaps every two hours) and taking corrective action. The advantage of using SAP ERP to manage credit is that the process is automated and the data are available in real time. The user can double-click the sales order to see company information, such as contacts, or to see payment history.

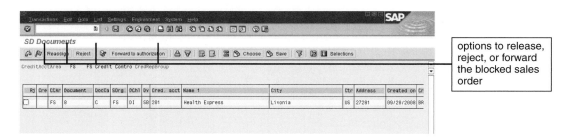

options to release, reject, or forward the blocked sales order

FIGURE 5-7   Blocked sales order

In the case of Fitter Snacker, the sales order clerk must manually check credit. If the clerk fails to do this, then a customer who is a bad risk may receive more credit. However, even when the clerk does perform the manual credit check, the credit decision can frequently be made in error, since the data are not current. Finally, because it is a manual system, blocked orders may erroneously become real orders, so sales may be affected. With the SAP ERP system, the check is automatic, the data are up-to-date, and it is a simple matter to review blocked sales orders.

Exercise 5.1

1. Create a document that describes Fitter Snacker's current credit management procedure. Write this document so that it could be used to train a new employee in the credit management process.
2. Revise the document to reflect the process improvements that would result if Fitter Snacker were performing credit checks using an integrated information system.

PRODUCT PROFITABILITY ANALYSIS

Business managers use accounting data to perform profitability analyses of a company and its products. When data are inaccurate or incomplete, the analyses are flawed. There are three main reasons for inaccurate or incomplete data: inconsistent recordkeeping, inaccurate inventory costing systems, and problems consolidating data from subsidiaries. In the next section, we'll look at each of these causes.

Inconsistent Recordkeeping

Each of FS's marketing divisions maintains its own records and keeps track of sales data differently. When the Direct Sales Division records a sale, the files include a code for a sales region (Northeast, Southeast, and so on). When the Distributor Division records a sale, the files include a code for the state. Suppose that an FS executive asks for a report that summarizes monthly sales dollars for all states in the Mid-Atlantic region for the previous year. Neither division's records are set up to answer that question. An FS accountant would need to perform this tedious series of steps:

1. Go to the source sales documents.
2. Code each document by state and region (as the case may be).
3. Summarize the data by state and region to produce the report. This would have to be done by hand, entering the data into a spreadsheet for review, or by some other means.

Now, suppose FS's management wants to evaluate the efficiency of Production's operations. Production uses paper records, so data must be taken from the paper records and entered into a spreadsheet. As often happens, those paper records might be inaccurate or missing, making the validity of the final report questionable.

There are many variations on this theme. Conceivably, a company's divisions do maintain the same data about a function, but often each division's system was created at a different time, each using a different file system. To answer a manager's question, at least one set of data would need to be rekeyed into a spreadsheet (or some other middleware program) for the merged analysis. Again, it is possible to get an answer, but doing so takes time.

Without integrated information systems, much of the effort of accounting and reporting to management resembles the situation in these examples: working around the limitations of the information systems to produce useful output.

With an ERP system, however, this sort of effort is minimized or eliminated because both divisions record and store their data in the same way, in the same records. If the company's process was changed to fit the best practices of the software when the system was installed, the managers of each division would have agreed on the way to store and collect data, as part of the system's configuration. Later, questions can be answered in a few minutes by any accountant (or manager or salesperson, for that matter) who understands how to execute a query in the database language, or how to use built-in management-reporting tools.

Inaccurate Inventory Costing Systems

Correctly calculating inventory costs is one of the most important and challenging accounting tasks in any manufacturing company. Managers need to know how much it costs to make individual products, so they can identify which products are profitable and which are not.

First, we will provide an overview of inventory cost accounting. Next, we will see how an ERP system can improve the accuracy of inventory cost accounting. Finally, we will discuss the rationale behind activity-based costing as a method to further improve the accuracy of inventory cost accounting.

Inventory Cost Accounting Background

A manufactured item's cost has three elements: the cost of raw materials, the cost of labor employed directly in the production of the item, and all other costs, which are commonly called **overhead**. Overhead costs include factory utilities, general factory labor (such as custodians or security guards), managers' salaries, storage, insurance, and other manufacturing-related costs.

Materials and labor are often called **direct costs** because the constituent amounts of each in a finished product can be estimated fairly accurately. On the other hand, the overhead items, called **indirect costs**, are difficult to associate with a specific product or a batch of specific products. In other words, the cause-and-effect relationship between an overhead cost (such as the cost of heat and light) and making a particular product (NRG-A bars) is difficult to establish.

Nevertheless, overhead costs are part of making products, so companies must have some way to allocate these indirect costs to items made. A common method is to use total machine hours, on the assumption that overhead is incurred to run the machines that make the products. Overhead costs are added up and then divided by the expected total machine hours to get an amount per hour. This value is then used to allocate overhead costs to products. If, for example, FS's overhead per machine hour is $1,000, and 10,000 bars are made in an hour, then each bar made would be allocated $0.10 of overhead ($1,000 ÷ 10,000). Another allocation method distributes total overhead costs by direct labor hours, on the assumption that overhead costs are incurred so that workers can do their jobs.

Companies such as FS that produce goods for inventory typically record the cost of manufacturing during a period using a **standard cost**. Standard costs for a product are established by studying historical direct and indirect cost patterns in a company and taking into account the effects of current manufacturing changes. At the end of an accounting period, if actual costs differ from standard costs, adjustments to the accounts must be made to show the cost of inventory owned on the balance sheet and the cost of inventory sold on the income statement.

For example, FS might determine that each NRG-A bar should cost $0.75 to make—that is, the cost of raw materials, labor, and overhead should equal $0.75, given the budgeted number of units. That amount would be FS's standard cost for a bar. During a month, FS might make 1 million NRG-A bars. Using the standard cost, it would increase its balance sheet inventory account by $750,000. Also, assume that the company sells 800,000 bars in the month. In the income statement, the cost of the sales would be shown as $600,000. The inventory account would be reduced by $600,000, because the company no longer has the units to sell.

If actual costs in the month equaled standard costs, no balance sheet or income statement adjustments would be needed. Actual costs never exactly equal expected costs, however, so adjustments are needed. The differences between actual costs and standard costs are called **cost variances**. Note that cost variances arise with both direct and indirect costs. These variances are calculated by comparing actual expenses for material, labor, utilities, rent, and so on, with predicted standard costs.

If the company keeps records for the various elements separately, compiling variance adjustments can be quite tedious and difficult. If products are made by assembling parts that are made at different manufacturing sites, and the sites use different information systems, the adjustments may be very imprecise.

ERP and Inventory Cost Accounting

Many companies with unintegrated accounting systems analyze their cost variances infrequently. Often, these companies do not know how much it actually costs to produce a unit of a product. As the following example illustrates, knowing precisely how much production costs can be very important.

Suppose FS has an opportunity to sell 300,000 NRG-A bars to a new customer. This is a huge order for FS. The customer wants a price of $0.90 per bar. FS's standard cost per bar is currently $0.70, based on information that is two months old. FS knows that the costs to manufacture snack bars have been increasing significantly in the past months. FS does not want to sell at a loss per unit, but it also does not want to lose a large order or a potentially good customer. Because of the difficulty of compiling all the data to calculate cost variances, FS only analyzes cost variances quarterly, and new data will not be available for another month. Should FS accept the large order?

If FS had an ERP system, employees throughout the company would have recorded costs in the company-wide database as they occurred. The methods for allocating costs to products and for computing variances would have been built into the system when it was configured. Thus, the system could compute variances automatically when needed. Not only would this simplify the process of adjusting accounts, FS's management would always have accurate, up-to-date information on cost variances. FS could decide whether it can profitably sell snack bars for $0.90 each. Furthermore, with a properly operating sales and operations planning process, FS could determine whether it has the capacity to complete the order on time, as well! If overtime would be required to complete the order, then analysts could use the planning capabilities of the ERP system to evaluate costs using overtime production.

ERP system configurations allow analysts to track costs using many bases—by job, by work area, or by production activity. This means that unit costs can be computed using different overhead allocation bases, allowing an analyst to play "what if" with product profitability decisions. In an unintegrated system, doing such multifaceted tracking would be time-consuming and difficult.

Product Costing Example

Suppose Fitter Snacker wishes to update standard costs for NRG-A bars. By analyzing the company's recent indirect costs related to the products produced, FS's cost accountants have calculated new overhead rates. Because the material costs are much larger than direct labor costs at Fitter Snacker, the company applies production overhead as a percentage of direct material costs. The new rate for production overhead is 100 percent of direct material costs.

Figure 5-8 shows the product cost analysis for the NRG-A bar. The cost analysis is based on seven cases of bars. The recipe for a 500-pound batch of NRG-A bars was given in Chapter 4, Figure 4-16. This information is repeated in the cost analysis of Figure 5-8, along with the cost of each of these materials. Multiplying the quantity of materials by the unit cost and summing the results gives a direct material cost of $537.65. Applying the production overhead rate of 100 percent to this direct material cost gives a production overhead cost equal to the direct material cost, or $537.65. The direct labor cost to mix the dough and bake the snack bars is $54.50. It is important to note that the labor cost is only about 10 percent of the direct material cost, which is why the Fitter Snacker company has chosen to apply production overhead costs based on direct material only.

The sum of direct materials, production overhead, and direct labor is the cost of goods manufactured (COGM). Fitter Snacker currently uses a rate of 30 percent of the cost of goods manufactured to estimate the sales and administrative costs. Adding the sales and administrative costs to the COGM gives the cost of goods sold (COGS). Because the COGM and COGS were estimated based on the BOM (recipe) from Chapter 4 that produces seven cases of snack bars, the figures must be divided by seven to give the COGM and COGS on a per-case basis. Figure 5-8 shows that these are $161.40 and $209.82, respectively.

The product cost analysis lets you determine whether selling 300,000 NRG-A bars to a new customer for a price of $0.90 per bar would earn a profit for Fitter Snacker. Given that there are 24 bars in a box and 12 boxes in a case, the current cost for an NRG-A bar is:

$$\frac{\$209.82 \text{ / case}}{(24 \text{ bars/box})(12 \text{ boxes/case})} = \$0.72 \text{ / bar}$$

Fitter Snacker can sell the bars at $0.90 and make a profit of $0.18 per bar.

NRG-A Bar Product Cost Analysis (7 cases)				
Ingredient	Unit of measure	NRG-A	Cost per unit of measure	Direct Material Cost
Oats	lb	300	$0.20	$60.00
Wheat germ	lb	50	$0.30	$15.00
Cinnamon	lb	5	$3.00	$15.00
Nutmeg	lb	2	$4.50	$9.00
Cloves	lb	1	$5.50	$5.50
Honey	gal	10	$6.40	$64.00
Canola	gal	7	$1.70	$11.90
Vit./min. powder	lb	5	$18.45	$92.25
Carob chips	lb	50	$2.10	$105.00
Raisins	lb	50	$3.20	$160.00
Total Material				$537.65
Production of OH (100% of Direct Material)				$537.65
Direct Labor				54.50
Cost of Goods Manufactured				1,129.80
Sales and Administrative (30% of COGM)				338.94
COGS				1,468.74
COGM per case				$161.40
COGS per case				$209.82

FIGURE 5-8   Product cost analysis for NRG-A bar

Exercise 5.2

Given the following product costs for the Fitter Snacker company:

Protein Powder (lb)	$4.40
Hazelnuts (lb)	$1.64
Dates (lb)	$3.55

and the product information in Figure 4-16, estimate the COGM and COGS on a per-case basis for the NRG-B bar. Use the same direct labor costs and overhead percentages used for the NRG-A bar product cost analysis in Figure 5-8.

Product Cost Analysis in SAP ERP

Developing product costs in a large company can be a very time-consuming task. There may be thousands of complicated products, and the task of gathering the required information and insuring its accuracy can be a major challenge. An advantage of an integrated information system like SAP ERP is that timely, accurate information is available in the information system. The key information for a cost analysis is the direct material and direct labor. In SAP ERP, the direct material is determined from the bill of material (BOM), which is managed in the Production Planning (PP) module. Direct labor is determined from the product routing, which describes the machines and workcenters that are used in the production of a product, and is also stored in the PP module. The routing, combined with other information maintained in the PP module, allows the SAP ERP system to determine the quantities of direct material and direct labor used in a product. These production data, combined with material cost information stored in the FI module, provide the basis for a product cost analysis.

In the SAP ERP system, the method for developing a product cost is called a **product cost variant**. The product cost variant is basically the procedure for developing a product cost analysis. Once a product cost variant is developed, it only takes seconds for the SAP ERP system to gather the required information and create a product cost estimate. Figure 5-9 shows how the SAP ERP system provides a material cost estimate for the NRG-A bar. Not only does the SAP ERP product cost tool greatly reduce the time required to develop cost estimates, it also increases their accuracy because it gets its data directly from modules, where the data are real-time and maintained by the users.

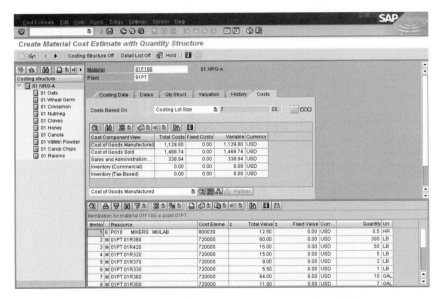

FIGURE 5-9   Material cost estimate for the NRG-A bar in SAP ERP

Activity-Based Costing and ERP

A trend in inventory cost accounting is toward **activity-based costing (ABC)**. In ABC, accountants identify activities associated with overhead cost generation, and then keep records on the costs *and* on the activities. The activities are viewed as causes (drivers) of the overhead costs. This view treats overhead costs as more direct than traditional cost-accounting methods have treated them. ABC tries to avoid rough allocation procedures in an attempt to assign costs more precisely to individual products. A company using ABC to provide more accurate cost allocations can determine which products have the highest profit margin, information that is crucial for making strategic decisions on product lines.

ABC is often used when competition is stiff, overhead costs are high, and products are diverse.

Consider this example from FS's operations. Suppose that storage of raw materials is considered an activity. Assume that storage activities differ between NRG-A and NRG-B bars because the ingredients are different, and that some of these storage activities are more labor-intensive than others. In an ERP system, FS would keep track of the various activities (how often they occur) and the cost of each. Later, when determining the profitability of each kind of bar, FS cost accountants would add in storage costs, based on the number of storage activities required by each type of bar. This costing is more precise than computing an average storage cost based on total storage costs and machine hours, and then allocating that amount to each kind of bar. Conceivably, if the activities differ enough from one bar to the next, one could be significantly more or less profitable than the other. This fact would be revealed by the ABC approach, but *not* by traditional cost-accounting approaches. Letting managers see that difference is the value of an information system that supports ABC.

Not all overhead costs can be linked to products by their activities. However, many can, depending on the company and the manufacturing situation. For many companies, the cost and effort required to implement ABC is justified by the value of the improved information yielded.

ABC requires more bookkeeping than traditional costing methods because a company must do ABC in addition to traditional costing, and because ABC requires a company to keep track of instances of activities, not just the costs. Companies often use ABC for strategic purposes, and at the same time use traditional costing for generally accepted accounting principles such as bookkeeping and taxes. Having an integrated information system allows a company to do both kinds of accounting much more easily. A recent study of companies with and without ERP revealed that: (1) ERP companies had nearly twice as many cost-allocation bases to use in management decision making, and (2) the ERP companies' managers rated their cost-accounting system much higher. Companies with ERP systems value their cost-accounting systems more than companies without ERP do, and ERP companies also have more faith in the numbers from their system.

Companies with Subsidiaries

Some companies have special operations that make closing their books at the end of an accounting period a challenge. Companies that have subsidiaries or branches face such a challenge. Because company managers want the big picture of overall operations and profitability, account balances for each entity must be compiled and forwarded to the home office. A consolidated statement for the company as a whole must then be created.

You might think this would be merely an arithmetic problem: add up cash for all the entities, accounts receivables for all the entities, and so on through the accounts. The job, however, is more difficult than that. Problems can arise from two sources: accounts stated in another country's currency must be converted to U.S. dollars (in the case of a U.S. parent company), and transactions between companies and their subsidiaries must be eliminated from the accounts.

Currency Translation

The following scenario illustrates the problems of **currency translation**. Assume one euro is worth $1.25. A company's European subsidiary reports cash of 1 million euros at the end of the year. When the European subsidiary's balances are consolidated with those of the U.S. parent company at the end of the year, $1,250,000 will be recorded. The same sort of translation would be done for all the European subsidiary's accounts.

A complicating factor is that exchange rates fluctuate daily. An ERP system can be configured to access daily exchange rates and translate daily transactions automatically.

ANOTHER LOOK

SAP in Use at Basell

Author Ellen Monk had a chance to interview Maureen Sullivan, an accounting manager at Basell, a privately held chemical company based in the Netherlands. Basell is the world's largest producer of polypropylene and the largest producer of polyethylene in Europe. Both products are used for making plastics for various industries such as automotive, packaging, and toys. Basell's customers also include soft drink companies and consumer products companies.

EM: When did your company implement SAP?

MS: We've had SAP for about 10 years. After the initial implementation, we've had some upgrades and also one downgrade. The downgrade was from version 4.6 to 4.5 because of a merger, where we had to use the same version as the acquiring company. We have many modules within SAP, for example, MM, FI, CO, Fixed Assets, and PP. Company-wide we consolidate with Hyperion, not with SAP. [Consolidation is the combining of financial statements of the parent company and subsidiaries.]

EM: How does your company handle exchange rates in the SAP system?

MS: There is an exchange rate table in the SAP system and the data in that table is changed daily at the company headquarters in the Netherlands. For accounting purposes at Basell North America, the month-end rate is used for the balance sheet reporting. There is a function in SAP FI to do that.

EM: Your company has hundreds of products. How does your company do product costing?

MS: Costing of raw materials and finished goods is done in the SAP system. Because we make polypropylene, there is not a cost associated with work-in-progress. The main raw material is valued at standard cost price, whereas other raw material is valued at moving average. In moving average, the inventory is valued at the cost price of the item. The finished goods at the producing location are valued at standard cost, but once those goods are moved to another location they are valued at moving average. The month-end costing utilizes the FIFO method. All overhead is included in the product cost. [FIFO stands for first in, first out. It's a way of costing goods such that the first goods acquired are the first costs charged to expense.]

continued

Intercompany Transactions

Transactions that occur between companies and their subsidiaries, known as **intercompany transactions**, must be eliminated from the books of the parent company because the transaction does not represent any transfer of funds into or out of the company.

Suppose that Acme Inc. owns Bennett Manufacturing. Bennett sells raw materials to Acme for $1 million. Acme then uses the materials to make its product. Bennett's sale is Acme's cost of sales. From the point of view of an outsider, money has merely passed from one part of the consolidated company to another. A company cannot make a profit by selling to itself.

Companies often do business with their subsidiaries. If a company does so, then such transactions will occur frequently. Keeping track of them and making the adjustments can be a challenge for the accountants.

ANOTHER LOOK

Integrated Accounting at NB Power

In recent years, the Canadian government has made some changes to its energy market. As a result, NB Power, the electric utility company in New Brunswick, was divided into a holding company and four operating entities—Generation, Nuclear, Transmission, and Distribution and Customer Service—which were simply cost centers within NB Power prior to 2002. The company split the responsibility for managing revenues, expenses, and financial statements among the four operating companies. Intercompany transactions were a huge chore because the budgeting and forecasting were done outside the SAP system. Now, each unit handles budgeting and forecasting within SAP. All intercompany data are handled within the SAP system, which makes forecasting and other accounting chores much easier.

Question:

1. How does an intercompany transaction differ from an intracompany transaction in the field of accounting?

MANAGEMENT REPORTING WITH ERP SYSTEMS

The integrated nature of an ERP system and the use of a common database and built-in management reporting create numerous benefits. Although reporting accounting information is commonplace, it is often very challenging for companies to generate the right reports for the right situation. Without an ERP system, the job of tracking all the numbers that need to go into a report is a monumental undertaking. With an ERP system, a vast amount of information is available for reporting purposes. Often companies take years after ERP implementation to figure out which reports are the most critical for decision making. In this section, we will examine some management-reporting and analysis tools available with ERP systems.

Document Flow for Customer Service

As you have seen, when an ERP system is used, all transactions in all areas of a company get posted in a centralized database. It is worth reemphasizing that the database *is* the company's "books." There is no separate set of books for Marketing or Production or Purchasing.

Thus, even though it is common usage to refer to "data flows" in an ERP system, it is actually a misnomer. Data do not flow from one ERP module to another because they are all in one place—the database. Each area views the same records. It might be better to speak of "data access" than of "data flows" when talking about how these areas use the common database. However, typical usage by companies (*data flows* and *data sharing*) is probably too ingrained to avoid.

As you learned in Chapter 3, each transaction that is posted in SAP ERP gets its own unique document number. This number allows quick access to the data. If you need to look up a transaction online, you do so by referencing the document number, which acts as an index to the appropriate database table entries.

In SAP ERP, document numbers for related transactions are associated in the database. This provides an electronic audit trail for analysts trying to determine the status of an order. The best example of this concept is the linkage of document numbers for a sales order. Figure 5-10 illustrates the document flow concept.

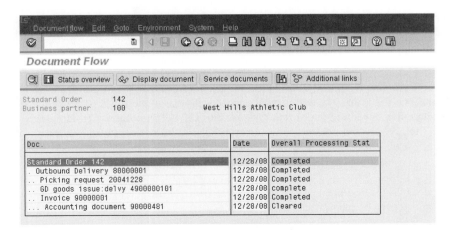

FIGURE 5-10 Document flow of a transaction in SAP ERP

The linked events shown in Figure 5-10 progress as follows:

1. When the order was placed, sales order document 142 was created.
2. The system recorded the delivery, which is the transfer of the order's requirements to the Materials Management module. It is denoted by document 80000001.
3. The picking request, which is the document that tells warehouse personnel which items make up the order, was created on Dec. 28, 2008, and given document number 20041228.
4. The goods were removed from inventory on the same day, an event recorded by document number 4900000101.
5. After the goods were issued, an invoice was generated so that the customer would be billed. The invoice was given document number 90000001.
6. At the same time, the accounting entries for the sale were generated. The posting document number is 9000481.

The document flow can be used to drill down to see the details of any one of these events. The term **drill down** refers to the ability to view the details behind a summary of information. For example, the user can double-click the order number (142) and see the details of the order—products ordered, quantity, customer name, and so on. From that display, the user can double-click the product numbers or the customer number to see details about them. To see the debits and credits in the accounting entry, the user can double-click 9000482 to see the scheduled entries.

Users can access the document flow from any SAP screen. If a customer were to call and ask about the status of an order, the clerk could access the document flow and see whether the goods had been shipped. If the customer called with an invoice number and questions about the billing, the customer service representative could use the document flow to access the appropriate documents in the chain of events, such as the original order or the picking request. This sort of research can be done quickly with SAP ERP. With unintegrated systems, establishing the audit trail and researching source documents can be very difficult and time-consuming.

Built-In Management-Reporting and Analysis Tools

Accounting records are maintained in the common database. The advantage of using a database is the ability to query the records to produce standard reports as well as answer ad hoc questions. An ad hoc question is one that is spontaneous. For example, a Fitter Snacker manager might run into an analyst's office and ask for an immediate sales report for the third quarter, snack bar division, by product. Traditional accounting packages are not optimized to set up and execute queries against accounting records, but database packages are. Therefore, when the records are kept in a database, the user gets a double benefit. The records can be kept in an accounting package, *and* the records can be queried because of the built-in database language.

Thus, a user who wants to identify the 10 largest orders placed by Health Express in the past year can execute a query to show the answer. In principle, this query could directly access the transaction records to get the answer. In practice, this would mean that analysts running queries would be accessing the records at the same time as current transactions are being recorded. This competition can slow down processing in even a large database, such as those used by ERP packages.

SAP's solution to this problem is to provide a **data warehouse** within each major module. A data warehouse is a repository for data from various sources. Analysts can use it without affecting the underlying data or performance of the transactional database. Users query the warehouse rather than the transaction database. For example, SAP ERP provides the Sales Information System (SIS) tool for analysts querying the sales records, and the Logistics Information System (LIS) tool for analysts querying the logistics (shipping) records. Both the SIS and the LIS come embedded with SAP ERP. Also, as mentioned in Chapter 2, SAP also sells its Business Warehouse (BW) product, a completely separate information system that extracts data from the SAP ERP system. With BW, users have great flexibility in defining data reports and analyses in a system that does not compete for system resources with transaction processes.

As the previous sections illustrate, an ERP system is a key component in creating management reports. Managerial accounting reports help the company's managers understand how the company is making money—what products are profitable, where costs may need to be reduced, and so on. Financial accounting reports are used to inform external parties—shareholders and government agencies—how well the company is doing financially. The importance of accurate accounting reports cannot be overstated, as you will see in the next section.

THE ENRON COLLAPSE

On October 16, 2001, Enron, then one of the world's largest electricity and natural gas traders, reported a $618 million third-quarter loss and disclosed a $1.2 billion reduction in shareholder equity, partly related to partnerships run by its chief financial officer (CFO), Andrew Fastow. Until that time, Enron had been a rapidly growing firm that was revolutionizing the energy business and making millionaires out of its investors. CEO Jeffrey Skilling, who resigned on August 14, 2001, for personal reasons, had helped transform the company from a natural gas pipeline company to a global marketer and trader of energy. The company had encouraged its employees to invest large portions of their 401K retirement savings accounts in Enron stock by matching employee contributions. On October 17, the day after Enron reported its tremendous third-quarter loss, the U.S. Securities and Exchange Commission (SEC) sent a letter to Enron asking for information about the loss. The SEC is dedicated to protecting investors and maintaining the integrity of the securities markets. Enron's high-flying business practices immediately began unraveling.

On October 22, 2001, Enron acknowledged an SEC inquiry into a possible conflict of interest related to the company's dealings with the partnerships run by CFO Fastow. Shares of Enron sank more than 20 percent on the news. Two days later, Enron ousted CFO Fastow. On November 8, Arthur Andersen, Enron's auditing firm, received a federal subpoena for documents related to Enron, and on December 2, Enron made the largest Chapter 11 bankruptcy protection filing in history. Clearly, the accounting records made public from Enron, which were released quarterly and were not audited, did not reflect the financial health of the company.

Enron began as an oil pipeline company in Houston in 1985. With the deregulation of electrical power markets, Enron expanded into an energy broker, trading electricity and other commodities. Unlike a traditional exchange, which brings together buyers and sellers, Enron entered into separate contracts with sellers and buyers, making money on the difference between the selling price and the buying price. Because Enron kept its books private, it was the only party that knew both prices. Enron's business developed over time, extending to practices that allowed firms to insure themselves against a range of risk factors, including changes in interest rates, weather, and a customer's inability to pay. The volume of these financial contracts was far greater than the volume of contracts to actually deliver commodities.

To manage the risk in these contracts, Enron was employing Ph.D.s in mathematics, physics, and economics. Risk management balances the opportunities offered by a business against the risks inherent in taking that business. As Enron's services became more complex and its stock soared, Fastow created partnerships between Enron and companies involved in Internet broadband technologies, computer technology, and energy, to name a few. Some of the partnerships were faked to mask billions of dollars in debt, allowing managers to shift debt off the books.

The partnerships that Fastow engineered were the subject of discussion well before Enron's bankruptcy. In a June 1, 1999, article in *CFO* magazine, Ronald Fink noted that the Financial Accounting Standards Board was looking at rule changes that would affect companies using creative financing techniques, like Enron. Enron owned a number of subsidiaries, but made sure that it owned no more than 50 percent of the voting stock. As a result, Enron was able to keep the debt and assets of these subsidiaries off Enron's own books. If Enron had not been able to use these creative accounting practices, the company would have had to report a much higher percentage of debt, which would have increased the costs that Enron paid to borrow money.

For years, Enron's financial statements had been audited by Arthur Andersen, a highly regarded accounting firm. As Enron's auditor, Andersen issued annual reports attesting to the validity of Enron's financial statements; it was supposed to function as an unbiased, incorruptible observer and reporter. Enron's October 16, 2001, press release characterized numerous charges against income for the third quarter as "non-recurring," even though Andersen believed the company did not have a basis for concluding that the charges would in fact be non-recurring. Indeed, Andersen advised Enron against using that term, and documented its objections internally in the event of litigation, but did not report its objections or otherwise take steps to correct the public statement.

Perhaps the most damning part of Andersen's indictment was the destruction of documents. On October 22, 2001, Enron acknowledged the Securities and Exchange Commission inquiry, and on October 23, Andersen personnel were called to urgent and mandatory meetings. Andersen employees on the Enron engagement team were instructed by Andersen partners and others to immediately destroy documentation relating to Enron, and were told to work overtime if necessary to accomplish the destruction. During the next few weeks, an unparalleled initiative led to the shredding of paper documentation and the deletion of hundreds of computer files.

Outcome of the Enron Scandal

The effects of the Enron scandal go both deep within and well beyond the company. Enron is approximately $63 billion in debt, but most of its more than 20,000 creditors will receive about one-fifth of the amount they are owed. Many of Enron's shareholders were Enron employees who invested their 401K accounts in Enron stock. Shareholders lost an estimated $40 billion dollars—in many cases, these individuals lost their entire life savings. A class-action lawsuit against financial institutions that had dealings with Enron (including Canadian Imperial Bank of Commerce, JPMorgan, and Citigroup) has produced more than $7 billion in settlements, although legal fees will consume a significant portion of this.

Thousands of workers lost their jobs, and 31 individuals were either charged or pled guilty to criminal charges. J. Clifford Baxter resigned as Enron vice chairman on May 2, 2001. He was found shot to death in his car on January 15, 2002, in an apparent suicide. Andrew Fastow, Enron's former chief financial officer, received a six-year prison sentence. His sentence had been limited to no more than 10 years as part of a plea agreement to testify against former CEO Jeffrey Skilling and CEO Ken Lay. Fastow had also involved his wife, Lea, in his crimes. She served a year in prison and a halfway house. Jeffrey Skilling was convicted on 19 counts of conspiracy, fraud, and insider trading in October 2006. He was ordered to pay nearly $45 million into a restitution fund for Enron's victims, and was sentenced to 24 years in jail. Although he is appealing his conviction, he began serving his sentence in late 2006. Ken Lay was convicted on fraud and conspiracy charges in May 2006, but two months later, prior to being sentenced, he died of a heart attack.

On June 15, 2002, jurors convicted the accounting firm Arthur Andersen for obstructing justice by destroying Enron documents while on notice of a federal investigation. Andersen had claimed that the documents were destroyed as part of its housekeeping duties, and not as a ruse to keep Enron documents away from the regulators. That October, U.S. District Judge Melinda Harmon sentenced Andersen to the maximum: a $500,000 fine and five years' probation. Those events were anticlimactic, however, as the former auditing giant had been all but dismantled by then. Once a world-class firm with 28,000 employees in the United States alone, Andersen has since been whittled down to about 200 people, most of them dealing with litigation and running a training center outside Chicago.

As a result of the failure of Enron, as well as the high-profile bankruptcies of World-Com and Global Crossing, the U.S. Congress passed the Sarbanes-Oxley Act of 2002. This act was designed to prevent the kind of fraud and abuse that led to the Enron downfall.

Key Features of the Sarbanes-Oxley Act

The Sarbanes-Oxley Act is designed to encourage top management accountability in firms that are publicly traded in the United States. Frequently, top executives involved in corporate scandals claim that they were unaware of abuses occurring at their company. Title IX of the Sarbanes-Oxley Act adds the requirement that financial statements filed with the Securities and Exchange Commission must include a statement signed by the chief executive officer and chief financial officer, certifying that the financial statement complies with the SEC rules. Specifically, the statements certify that "the information contained in the periodic report fairly presents, in all material respects, the financial condition and results of operations of the issuer." Anyone "willfully certifying any statement . . . knowing that the periodic report accompanying the statement does not comport with all the requirements set forth in this section shall be fined not more than $5,000,000, or imprisoned not more than 20 years, or both."

Title II of the act addresses auditor independence. Among other things, this section of the act limits the non-audit services that an auditor can provide. Among the non-audit services prohibited by the act are:

- Bookkeeping or other services related to the accounting records or financial statements
- Financial information systems design and implementation services
- Legal services
- Expert services unrelated to the audit
- Management functions
- Human resources functions
- Any other service that the Public Company Accounting Oversight Board (PCAOB) determines to be impermissible. The PCAOB was created in Title I of the act with broad powers to regulate audits and auditors of public companies.

Title IV of the act, Enhanced Financial Disclosures, specifies more stringent requirements for financial reporting. Section 404 of Title IV requires that a public company's annual report contain management's internal control report. This control report outlines management's responsibility for establishing and maintaining adequate internal control over financial reporting, and assesses the effectiveness of its internal control over financial reporting. Section 409 of Title IV addresses the timeliness of reports, and may require companies to file an SEC report within two days of a significant trigger event—for example, completion of an acquisition or default by a major customer.

IMPLICATIONS OF THE SARBANES-OXLEY ACT FOR ERP SYSTEMS

Certainly the Sarbanes-Oxley Act has significant implications for a firm's information systems. To meet the internal control report requirement, a company must first document the controls that are in place and then verify that they are not subject to error or manipulation.

An integrated information system provides the tools to implement internal controls, as long as the system is configured and managed correctly. However, even the passage of the Sarbanes-Oxley Act and the availability of state-of-the-art ERP technology cannot prevent insidious and systematic fraud similar to that of the Enron scandal. An ERP system relies on a central database with accurate information. ERP systems make it difficult to hide fraudulent dealings, and perhaps Enron's problems would have been more obvious to stakeholders of the company had the company implemented an ERP system. But it is unlikely that an ERP system can prevent all fraud.

On the positive side, companies with ERP systems in place will have an easier time complying with the Sarbanes-Oxley Act than will companies without ERP. Companies are also discovering that complying with the Sarbanes-Oxley Act will allow them to measure the success of their ERP system, and improve its performance, by revealing how powerful their ERP systems actually are. Most companies take years to find out how to use the full power of their ERP systems, and Sarbanes-Oxley provides them with a vehicle for exploiting those benefits.

ANOTHER LOOK

Sarbanes-Oxley Five Years Later

The Sarbanes-Oxley Act has been a part of doing business in the United States since 2002, and its impact is both widespread and unclear. A study by AMR Research has estimated the costs of compliance for U.S. business at $26 billion, while a controversial 2005 study conducted by Assistant Professor Ivy Zhang, then a graduate student, estimated the law's costs at a staggering $1.4 trillion. Other researchers claim the law has reduced investment in R&D and overall capital spending, and has led to an increase in the takeover of publicly traded companies by private-equity buyers (private firms do not have to comply with Sarbanes-Oxley requirements). No one argues that the law has been good for auditing firms, which is ironic given that the failure of auditors at Arthur Andersen contributed to the Enron collapse.

As firms have strengthened their financial controls, many have had to restate their financial results—1,403 firms did so in 2006. But as firms have improved in their ability to comply with the act, that number has been dropping.

The costs of complying with the act are also coming down. According to a 2007 survey of 200 companies with average revenues of $6.8 billion, the typical cost of Section 404 compliance was $2.9 million in 2006, which was 23 percent lower than in 2005. Relaxed audit standards approved by the SEC in 2007 will allow a more pragmatic, common-sense auditing approach that may lower compliance fees by up to 50 percent.

Question:

1. Use the Internet to research the latest regulations surrounding the Sarbanes-Oxley Act.

The next section explores ways in which SAP ERP and other ERP systems can prevent corporate fraud and abuse. Systems from vendors other than SAP have functionality similar to that of SAP ERP.

Archiving

One of the first things a new SAP ERP user notices is that the software offers very few ways to delete items. For example, the menus in the SAP ERP system related to material master data (master records that describe material characteristics) are shown in Figure 5-11. There are options for creating, changing, and displaying, but not simply deleting—the closest option is to flag for deletion. Before a material can be deleted from the SAP ERP system, a user must create an auditable record of its existence. Data are removed from the SAP ERP system only after they have been recorded to media (tape backup, DVD-R) for permanent storage. This permanent storage, or **archive**, allows auditors to reconstruct the company's financial position at any point in the past.

Suppose data could be freely deleted from the system. An unscrupulous employee could create a fictitious vendor, post an invoice from the vendor, have payment made for the bogus invoice to a Swiss bank account, and then delete all records of this transaction. It would be very hard to detect the fraud and probably impossible to find out who committed it, because the records would no longer exist.

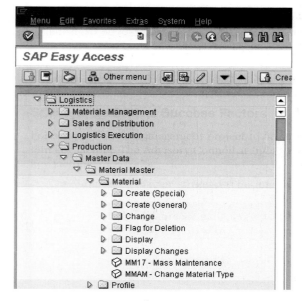

FIGURE 5-11   Transaction options for material master data

Not only does the SAP ERP system require archiving before data can be deleted, but it also keeps track of when data are created or changed. Figure 5-12 shows the Change Record for the material master. Each time a user changes the material master, the Change Record tracks the change in the data, who changed them, and when the change occurred.

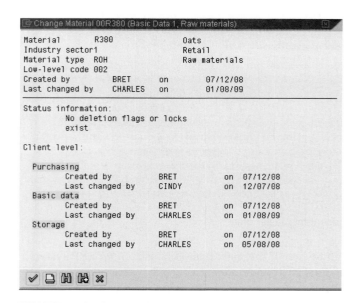

FIGURE 5-12   Change Record for material master

User Authorizations

Another way that an ERP system can prevent an unscrupulous employee from making payments to a fictitious vendor is through user authorizations and separation of duties. SAP ERP has sophisticated user administration tools that allow different levels of authorization management, to ensure that employees can perform only the transactions required for their jobs. One way that the SAP system controls user authorizations is through the Profile Generator, which provides a simple method for selecting the functions that a user should be allowed to perform.

Figure 5-13 shows a predefined role in the SAP ERP system for a user whose job involves managing material masters and bills of material. This employee can perform any transactions shown on the role menu in Figure 5-13.

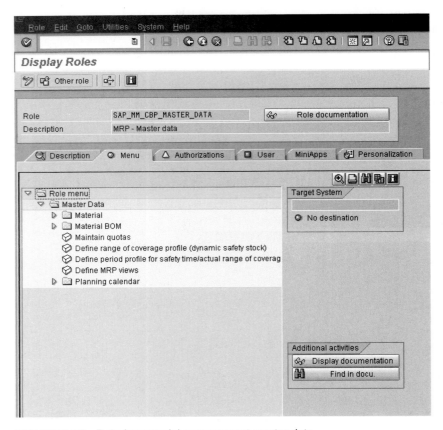

FIGURE 5-13   Role for material management master data

Managing authorized usage can be one of the most complex tasks in an ERP system. The joke in the world of authorizations is "If you're doing your job, we're not doing ours!" It may be a challenge to provide users with the proper authorizations in a timely manner, but most companies take the position that it is better to err on the side of caution, by taking a little longer to grant the right level of authorization, than to give users too much authority quickly.

Tolerance Groups

Another ERP way to make sure employees do not exceed their authority in financial transactions is to set limits on the size of transaction an employee can process. In the SAP ERP system, this is done using tolerance groups, as shown in Figure 5-14. As you learned in Chapter 2, tolerance groups are preset limits on an employee's ability to post transactions. Tolerance groups set limits on the dollar value for a single item in a document as well as the total value of the document. Just as importantly, they set a limit on payment differences. Suppose a customer has been invoiced for $1,005 but accidentally sends in a check for $1,000 to pay the invoice. The cost of requesting and processing a second payment for the $5.00 would cost both parties more than $5.00. In this case it is better to accept the $1,000 check as payment in full and account for the difference as a variance. In Figure 5-14, the Permitted payment differences section shows that the system would allow the user to process a payment that was in error by no more than 1 percent, or $10.00. Notice that the group field is blank in Figure 5-14. That means that this group is the default tolerance group. If an employee is not assigned to any other tolerance group, then by default the limits in the default group apply. As with authorizations, it is a safe policy to define a default tolerance group with low limits and err on the side of less authority rather than more authority.

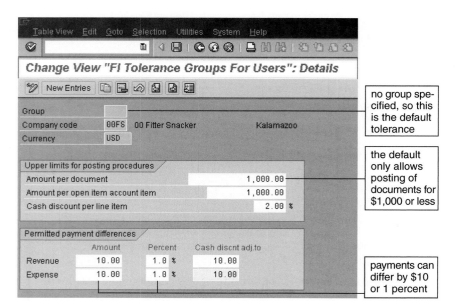

FIGURE 5-14   Default tolerance group

Financial Transparency

A key feature of any ERP system is the ability to drill down from a report to the source documents (transactions) that created it. For example, if sales figures for a region look unusually high, the user can double-click the figure in the report and drill down to review the specific sales orders that constitute the overall sales figures, to verify the results. The ability to drill down from reports to transactions makes it easier for auditors to confirm the integrity of the reports. Figure 5-15 shows a general ledger account balance for raw material consumption. This general ledger accounts for all raw material usage at the company.

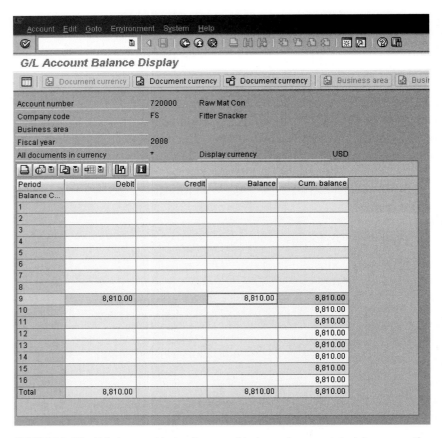

FIGURE 5-15   G/L (general ledger) account balance for raw material consumption

To see where the figures in this report come from, a user can double-click the line for Period 9, which brings up the screen in Figure 5-16. Figure 5-16 shows that two items make up the $8,810.00 raw material consumption. A manager might be intrigued by the $10 expense and want more information on what caused it. Clicking the Detail button provides more detailed information on the $10 expense, as shown in Figure 5-17. Fifty pounds of material were charged to cost center R010, which is a research and development cost center. From this screen even more details can be displayed.

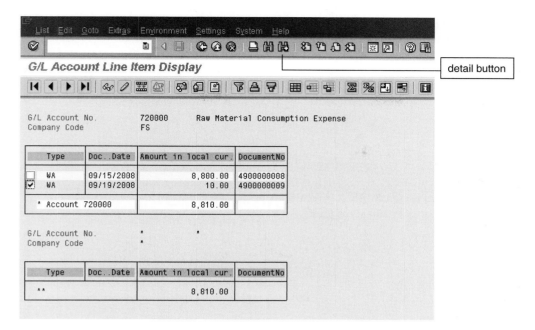

FIGURE 5-16 Documents that make up G/L account balance for raw material consumption

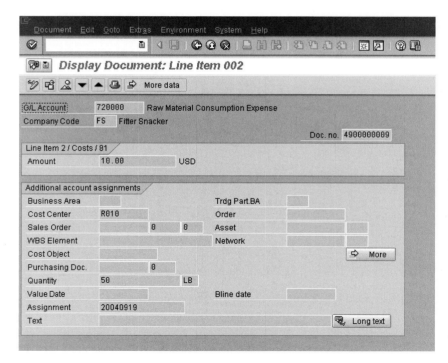

FIGURE 5-17 Details on $10.00 line item in G/L account for raw material consumption

With a few mouse clicks, an auditor can move from a summary statement of a general ledger account to find out all of the details related to an employee in the Research and Development department taking 50 pounds of oats from inventory for product development. With a properly configured and managed ERP system, there are direct links between the company's financial statements and the individual transactions that make up the statements, so that fraud and abuse can be detected more easily.

Exercise 5.3

Assume that you are the manager of the Accounting department at Fitter Snacker. Fitter Snacker still does not have an ERP system. What changes must you make in your accounting practices to prepare for the Sarbanes-Oxley Act? Write a proposal to your manager outlining your plan to be ready for the additional amount of reporting necessary. Remember that Fitter Snacker does not have an ERP system in place.

Chapter Summary

- Companies need accounting systems to record transactions and generate financial statements. The accounting system should let the user summarize data in meaningful ways. The data can then be used to assist managers in their day-to-day work and in long-range planning.

- With unintegrated information systems, accounting data might not be current, and this can cause problems for sales representatives trying to make operational decisions, such as granting credit. Data can also be inaccurate because of weaknesses in unintegrated systems, and this problem can affect decision making and therefore profitability.

- Closing the books at the end of an accounting period can be difficult with an unintegrated IS, but it is relatively easy with an integrated IS. Closing the books means zeroing out the temporary accounts.

- Using an integrated IS and a common database to record accounting data has important inventory cost-accounting benefits. More precise recordkeeping is possible, and this can lead to more accurate product cost calculations. These, in turn, can help managers determine which products are profitable and which are not.

- The use of an integrated system and a common database to record accounting data has important management-reporting benefits. The user has built-in drill-down and query tools available as a result.

- The introduction of the Sarbanes-Oxley Act, a 2002 U.S. federal regulation written and passed in the wake of the Enron collapse, promoted management accountability by requiring extra financial approval and reporting. Because ERP systems can help companies meet the requirements of this legislation, the act has increased the demand for integrated data reporting.

Key Terms

Accounts receivable	General ledger
Activity-based costing (ABC)	Income statement
Archive	Indirect costs
Balance sheet	Intercompany transaction
Cost variance	Managerial accounting
Currency translation	Product cost variant
Data warehouse	Profit and loss (P&L) statement
Direct costs	Overhead
Drill down	Standard costs
Financial accounting	

Exercises

1. Describe the differences between financial and managerial accounting. How can ERP systems benefit workers in both?

2. This exercise tests your understanding of the information needed to trace a sale through a multistep ERP sales cycle: sales order, inventory sourcing, delivery, billing, and payment. Assume that an order has been placed with your company's ERP system. These events occur:

   a. The system automatically checks the customer's credit and finds it to be acceptable. The order is recorded for the delivery date requested.

   b. The system schedules the production of the goods. (There is not enough inventory to ship from stock.)

   c. The system schedules raw material orders from the vendors to make the goods ordered.

   d. The raw materials are received and stored.

   e. The goods are produced and reserved for shipment to the customer.

   f. The system schedules the delivery, and the goods are put on the delivery truck. An invoice is printed and included with the shipment.

   g. Shipping notifies Accounting of the shipment's details.

   h. A month later, the customer sends in payment, which is recorded in accounting.

   For each of these events, list the information that must be recorded in the common database. You do not need to know how to use a database to do this, nor do you need to understand bookkeeping. At each step, did the wealth of the company increase or decrease? At each step, how did the company's obligations to outsiders change? At each step, how did the obligations of outsiders to the company change?

3. Suppose it is 4:00 p.m., September 29, 2008. Fitter Snacker's CEO sends this e-mail message to the accounting manager: "I need to meet with the Board of Directors tomorrow morning. The Board members are concerned about the current sales of our energy bars. They would like to see sales data for today, 9/29, as a typical day. Please complete the attached report and have it on my desk by 9:00 a.m. tomorrow, September 30." The blank table shown in Figure 5-18 is attached to the memo.

   Given the company's sales order data-processing practices, why won't the accountants be able to easily accomplish this task? Cite the practices that cause difficulty, and explain. If FS had an ERP system, why would this task be easy to complete? To answer this question, review Chapters 3 and 4 to see how sales orders are processed.

Sales data September 29	Distributor Division	Direct Sales Division	Total
# Bars sold			
NRG-A			
NRG-B			
Total			
$ Value of bars sold			
NRG-A			
NRG-B			
Total			
# Customers sold to			
NRG-A			
NRG-B			
Total			

FIGURE 5-18

4. The following exercise will test your understanding of FS's current credit-check system. In each situation, you are given background data and information about documents in the system.

SITUATION 1

Background data	
Today's date	6/29/09
Current list price, NRG-A	$1.50/bar
Current list price, NRG-B	$1.60/bar
Accounts receivable balance at start of business day, ABC Corp.	$9,000
Credit limit, ABC Corp.	$12,000
Current order	
Product	NRG-A
Amount	4 cases (1,152 bars)
Price	List
Ship to	ABC headquarters
Date desired	7/5/09
Next invoice number	A1001
Documents in system	
No documents relating to ABC are in the system.	

a. Given the state of the system, will credit be granted or denied for ABC Corporation's current order?

b. Suppose the system processed data in a more timely way. Should credit be granted or denied?

SITUATION 2

Background data	
Today's date	7/3/09
Current list price, NRG-A	$1.50/bar
Current list price, NRG-B	$1.60/bar
Accounts receivable balance at start of business day, KLM Corp.	$6,000
Credit limit, KLM Corp.	$8,000
Current order	
Product	NRG-B
Amount	5 cases (1,440 bars)
Price	List
Ship to	KLM headquarters
Date desired	7/8/09
Next invoice number	A1200
Documents in system	

Purchase order KLM 82332 for three cases (864 bars) of NRG-A. This order is in the sales order entry program, but it has not been transferred to the accounting program (thus, Accounting does not yet know about this sale).

c. Given the state of the system, will credit be granted or denied for KLM Corporation's current order?

d. If the system processed data in a more timely way, would credit be granted or denied?

SITUATION 3

Background data	
Today's date	7/13/09
Current list price, NRG-A	$1.50/bar
Current list price, NRG-B	$1.60/bar
Accounts receivable balance at start of business day, ACORN Corp.	$6,000
Credit limit, ACORN Corp.	$6,000
Current order	
Product	NRG-A
Amount	150 boxes (3,600 bars)
Price	List
Ship to	ACORN headquarters
Date desired	7/15/09
Next invoice number	A1300
Documents in system	

A check from ACORN was received in yesterday's mail and entered into the accounting system. The check is for $2,000, applied to in voices from June's sales. The sales clerks are working from credit-limit printouts prepared at the beginning of the week (two days ago).

c. Given the state of the system, will credit be granted or denied for ACORN Corporation's current order?

d. If the system processed data in a more timely way, would credit be granted or denied?

5. ERP systems save time for accountants in many ways. Some researchers expected that employment of accountants would diminish with the implementation of an ERP system. In reality, accountants are needed more than ever in industry. Explain how the Sarbanes-Oxley Act has impacted the demand for accountants. Research the job market for accountants to prove this point.

6. Companies usually prepare division- and company-wide budgets each year. These budgets show important monthly results that the company plans to achieve: sales, cost of sales, inventory levels, cash on hand, and other key data. Such budgets are effective for planning and controlling operations if they are designed as "flexible" budgets. A flexible budget is restated as conditions change from month to month, so goals remain reasonable and useful for evaluating performance. If information systems are unintegrated, getting data from the company's departments, in order to create the budget, is a chore. Keeping the data current is very difficult and often is not done. If not kept up to date, budgets are not useful for planning and controlling operations. Thus, the flexible budget concept is a good idea, but it is difficult to achieve. Because ERP makes flexible budgeting more achievable, ERP helps management discharge its planning and controlling roles better. Why do you think flexible budgeting would be more achievable with an ERP system? List your reasons, and explain.

7. Review FS's unintegrated production and purchasing procedures, described in Chapter 4. How would its current job-scheduling, production, and purchasing procedures result in variances from standard costs? Why would FS have trouble researching these costs at month's end to adjust the "standard costs per unit" to accurate "actual costs per unit"?

For Further Study and Research

Barnhart, Todd M. "The Financial Supply Chain." *Darwin Magazine*, April 2004.

Bloomberg News. "Lawyer who negotiated Enron settlements retiring." *Bloomberg News*, August 28, 2007.

Economist.com. "Sarbanes-Oxley: Five years under the thumb." *The Economist*, July 26, 2007. http://www.economist.com/business/displaystory.cfm?story_id=9545905&CFID=19771870&CFTOKEN=7941782.

Fink, Ronald. "Balancing Act: Will a new accounting rule aimed at off-balance-sheet financing trip up Enron?" *CFO*, July 1, 1999.

Flood, Mary. "Andersen conviction affirmed: Appeals court upholds verdict in Enron case." *Houston Chronicle*, July 1, 2004.

Goff, John. "They Might Be Giants: It's been nearly two years since Arthur Andersen went under and Sarbanes-Oxley was passed. Have the Big Four audit firms changed since then?" *CFO*, January 12, 2004.

Hays, Kristin. "Ex-Enron CFO Fastow sentenced to 6 years in prison." *Houston Chronicle*, September 26, 2006.

Johnson, Carrie. "Enron's Lay Dies of Heart Attack: Convicted Founder Faced Life in Prison." *Washington Post*, July 6, 2006.

KPMG LLP. "Sarbanes-Oxley Section 404: Management Assessment of Internal Control and the Proposed Auditing Standards." White Paper, March 2003. http://www.kpmg.ca/en/services/audit/documents/SO404.pdf.

KPMG LLP. "Sarbanes-Oxley: A Closer Look." White Paper, January 2003.

Krumweide, Kip R., and Win G. Jordan. "Reaping the Promise of Enterprise Resource Systems." *Strategic Finance*, October 2000, 49–52.

Lubin, Joann S., and Kara Scannell, "Critics See Some Good from Sarbanes-Oxley." *The Wall Street Journal*, July 30, 2007.

McClenahen, John S. "The Book On The One-Day Close." *IndustryWeek.com*, April 1, 2002. http://www.industryweek.com/ReadArticle.aspx?ArticleID=1058.

Preacher, Debbie. "Sarbanes-Oxley: A Business Blessing in Disguise." ebizq.net, July 17, 2005. http://www.ebizq.net/topics/com_sec/features/6116.html.

SAP.com. "SAP Customer Success Story: NB Power: Deferred Restructuring Elicits Impressive Flexibility from SAP Consulting." 2004. http://www.sap.com/platform/netweaver/pdf/CS_NB_Power.pdf.

Sarbanes-Oxley Act, H.R. 3763, Title III, Section 302, §1350 (a)(3).

Sarbanes-Oxley Act, H.R. 3763, Title IX, Section 906, §1350 (c).

U.S. Securities and Exchange Commission. "The Investor's Advocate: How the SEC Protects Investors, Maintains Market Integrity, and Facilitates Capital Formation." http://www.sec.gov/about/whatwedo.shtml.

WashingtonPost.com. "Timeline of Enron's Collapse." *Washington Post*, July 9, 2004. http://www.washingtonpost.com/ac2/wp-dyn?pagename=article&node=&contentId=A25624-2002Jan10.

CHAPTER **6**

HUMAN RESOURCES PROCESSES WITH ERP

LEARNING OBJECTIVES

After completing this chapter, you will be able to:

- Explain why the Human Resources function is critical to the success of a company.
- Describe the key processes managed by a Human Resources department.
- Describe how an integrated information system can support effective Human Resources processes.

INTRODUCTION

A company's employees are its most important resources. The Human Resources (HR) department is responsible for many of the activities that a company performs to attract, hire, reward, train, and, occasionally, terminate employees. The decisions made in the HR department can affect every department in the company. Companies are increasingly aware of the importance of an experienced, well-trained workforce and have begun using the term **human capital management (HCM)** to describe the tasks associated with managing a company's workforce.

As a company grows from a small business to a large organization, the need for an organized and effective HR department becomes increasingly important. The responsibilities of an HR department usually include:

- Attracting, selecting, and hiring new employees using information from resumes, references, and the interview process

- Communicating information regarding new positions and hires throughout the organization and beyond

- Ensuring that employees have the proper education, training, and certification to successfully complete their duties

- Handling issues related to employee conduct

- Making sure employees understand the responsibilities of their jobs

- Using an effective process to review employee performance and determine salary increases and bonuses

- Managing the salary and benefits provided to each employee and confirming that the proper benefits are disbursed to new and current employees

- Communicating changes in salaries, benefits, or policies to employees

- Supporting management plans for changes in the organization (expansion, retirements, and so on) so that competent employees are available to support business processes

Making sure that these tasks are accomplished and that valid information is communicated requires an effective system to control the flow of information. In this chapter, we will explore the role of an integrated information system in Human Resources.

PROBLEMS WITH FITTER SNACKER'S HUMAN RESOURCES PROCESSES

As with Fitter Snacker's other processes, personnel management relies on paper records and a manual filing system. This setup creates problems because information is not readily accessible or easy to analyze. The HR department's recruiting, hiring, and post-placement processes would operate more efficiently with an integrated system.

Recruiting Process

Fitter Snacker has three employees in its HR department. Problems occur because of the large number of HR processes (from hiring and firing to managing health benefits), the lack of integration among all departments, and the number of people with whom HR interacts. Many of the HR problems also result from inaccurate, out-of-date, and inconsistent information.

For example, suppose a department has an opening for a new employee. The department supervisor communicates this need to the Human Resources department by filling out a paper job vacancy form that describes the position, lists the qualifications a candidate must have to fill the position, specifies the type of position (temporary, part-time, full-time, or co-op/intern), and states when the position will become available. Human Resources takes this information, verifies that the position needs to be filled, and gets final approval from the president of Fitter Snacker to begin the recruiting process. Because there is no central information system, the details on the job vacancy form are frequently inconsistent among, and sometimes within, departments.

Usually the job is first posted internally, so that current employees have the opportunity to apply for the position. If no current employees are acceptable for the position, then Fitter Snacker posts the position externally.

A number of problems can arise in the recruiting process. First, the description of the qualifications required for the job may be incomplete or inaccurate, sometimes because the supervisor is in a hurry, sometimes because the supervisor is not aware of all of the functions required for the position, and sometimes because the supervisor assumes that all candidates will have certain basic skills. Second, if the job vacancy form is lost or not routed properly, the Human Resources department will not know that the position is available, while the supervisor will assume that the paperwork is in process. When this happens, the department ends up shorthanded, creating tension or animosity between the departments. Obviously, this problem will occur more frequently when job openings are circulated by paper. With an integrated information system, the job information is available immediately and easier to monitor. Another serious recruiting problem related to a paper-based hiring process is the potential loss of a good candidate due to drawn-out hiring practices or lost data.

Although Fitter Snacker does not use recruiting agencies or Internet job sites such as Monster.com to find candidates, it does use several other methods to find people for its jobs. FS publishes its job vacancies on the company's Web site, in local newspapers, and, in the case of a professional position, in national publications. In addition, a representative from the HR department attends career fairs and recruits on college campuses for prospective candidates. Occasionally, referrals are made by other Fitter Snacker employees, and sometimes individuals searching for open positions at Fitter Snacker send unsolicited resumes.

Filing and keeping track of resumes and applications is a continuing challenge. Fitter Snacker has dozens of jobs with different titles and descriptions, and the company receives dozens of resumes and applications each day. The HR department must classify and file all applications and resumes according to the appropriate description. For example, if the resume of a mechanical engineer is accidentally filed with resumes of candidates applying for jobs in the accounting department, the mistake may not be discovered in time to include the engineer in the search process. In this case, Fitter Snacker may not hire the best person for the engineering position, and the mistake might damage Fitter Snacker's reputation.

Keeping the applicant's data on a paper form means that retrieving the applicant data and using it to evaluate candidates is also a challenging task. To generate a list of potential candidates, Human Resources evaluates the resumes and applications it receives in response to a job posting, and also reviews filed applications that are less than one year old. These resumes and applications must be photocopied and then circulated through the department making the job request. Frequently, more than one person in the requesting department reviews the applications, and because the applicant data are on paper, managers review the applicant files sequentially, slowing the review process.

The Interviewing and Hiring Process

At Fitter Snacker, the requesting department develops a **short list** of candidates for the position by selecting up to three applicants, based on the data provided by HR. Human Resources contacts the candidates on the short list, schedules interviews, and creates a file for each candidate. A candidate's file includes a form that shows when the application was received, the position(s) applied for by the candidate, and the date and time of any interviews. If this is the second time the candidate has applied for a job with the company, the form indicates the current status of the candidate: whether the candidate was interviewed and rejected, whether the candidate rejected a job offer, and so on.

If a candidate accepts the interview offer, the HR department makes the arrangements for the job candidate, including travel arrangements and a schedule of interview activities. A representative from the HR department conducts an interview that includes a discussion of the applicant's experience and questions relevant to the position for which the candidate has applied. The supervisor of the department in which the position exists also interviews the candidate, and other employees in the department are usually given time to talk to the candidate as well. For most professional positions at Fitter Snacker, the candidate is interviewed by the plant manager and, frequently, the company president.

After the initial interview process, HR updates the candidate's file to indicate whether he or she is still a possibility for hire. In some cases, a second interview is scheduled. Once HR has interviewed all the candidates on the short list, a representative of HR and the supervisor of the requesting department decide which candidates on the short list are acceptable, and rank them. If there is an acceptable candidate, the HR person makes the highest-ranking candidate a verbal job offer over the phone. If the candidate accepts the verbal offer, a written offer letter is sent, which the candidate must sign and return. Once the candidate formally accepts the written offer, his or her file is again updated, showing that the candidate has accepted the offer and will begin employment with the company on a specified date.

If there are no acceptable candidates from the short list, or if none of them accepts the job offer, then the process must be repeated, which at a minimum will require the development of a new short list but may involve starting over with a new job posting.

Many of Fitter Snacker's problems in the interviewing and hiring process have to do with information flow and communication. Fitter Snacker does not have group appointment calendar software, which would allow HR staff to easily find a time when all key personnel would be available to interview a candidate. A group appointment calendar (available in software packages such as SAP) allows users to check others' calendars in order to schedule meetings. Scheduling interviews is frequently a cumbersome process, requiring the Human Resources employee to coordinate the interview schedule between the candidate and the appropriate personnel at Fitter Snacker. Because this is done by e-mail and phone, it can take days and sometimes weeks to schedule an interview. A similar problem occurs after the interviews have been completed. Gathering feedback from all involved parties and ranking the candidates takes time and may require multiple meetings. Managing the travel arrangements and reimbursing candidates for their travel expenses are also cumbersome tasks. More than once, Fitter Snacker has lost a promising candidate to another company because of delays in the FS interviewing and hiring process.

After the candidate accepts the formal job offer, Fitter Snacker hires an HR consulting firm to perform a background check to verify that the candidate has not falsified any information and does not have a serious criminal record. Fitter Snacker outsources the background check because of the special skills required. If the background check is satisfactory, this information is also stored in the candidate's file, and the job offer stands. If the consultant finds evidence of falsified information or legal troubles, the file is likewise updated, and the job offer is rescinded with a written explanation.

After passing the background check, the new employee completes additional paperwork covering employment terms and conditions, tax withholding, and benefits. All Fitter Snacker employees must sign a form that states that the employee has been given a copy of—and agrees to abide by—the company's policies and procedures. The new employee must complete an IRS W-4 form, which tells the employer the correct amount of tax to withhold from the employee's paycheck. Next, the employee must attend an orientation session, during which HR personnel describe Fitter Snacker's benefits plan. Fitter Snacker offers a comprehensive benefits plan that gives employees a range of choices for healthcare plans, life insurance, retirement plans, and medical savings accounts. The employee's dependents may also be covered under Fitter Snacker's health insurance plan. If the employee elects dependent coverage, then HR must obtain basic information about each dependent to include in the employee's file.

Because employees must provide a significant amount of detailed data to properly manage compensation and benefits, it is not surprising that Fitter Snacker frequently has problems enrolling new employees in the correct benefits plans and establishing the proper payroll deductions. It can often take months to manage the new employee's compensation and benefits correctly. The enrollment issues can generate many time-consuming phone calls to HR management—calls that would not be needed with an integrated system.

ANOTHER LOOK

Challenges in Hiring Talent

Talented workers are hard to come by. The average quality of a worker has declined by 10 percent since 2004, and the time it takes to hire a talented worker has gone from 37 days to 51 days, so HR departments must be vigilant in their recruiting efforts. Using an integrated system, such as an ERP system, helps HR departments identify and retain that rare talent. Seventy-five percent of managers surveyed by the Corporate Executive Board said that attracting and retaining talented workers was their top goal.

The consulting firm McKinsey & Company groups jobs into three categories: transformational, transactional, and tacit. The first two job types—converting raw materials into products and performing easily automated business events—are shrinking compared to the tacit category, which requires a worker to have a high level of judgment. From 2000 to 2006, the number of jobs requiring a high level of judgment increased 2.5 times more than the transactional jobs, and three times compared with all jobs. Roughly 40 percent of U.S. employers now require workers with a high level of judgment. Furthermore, with baby boomers reaching retirement age, the consulting firm RHR International estimates that by 2012 the nation's 500 largest companies will lose half of their top managers.

Companies are trying to find ways to lure workers away from their current jobs. With downsizing a concern for many workers, employees are often eager to be lured away. To find candidates, companies will comb through lists of attendees at conferences, look for scientists who created new patents, or even buy information on the competition. Companies are also using corporate Web sites for recruiting. Surveys have shown that 95 percent of large companies in North America use their corporate Web sites for hiring. The sites often are connected to HCM software at the home office that screens resumes.

The hiring of good employees is so vital to a company's success that HR managers are now commanding larger salaries. HR executives for Black & Decker, Home Depot, Viacom, and Timberland are among those companies' five highest-paid employees.

Questions:

1. Why is the hiring process more crucial than ever for companies?
2. List some incentives, other than salary, that a company could use to encourage a prospective employee to accept a job offer.

Human Resources Duties After Hiring

The Human Resources department has responsibilities that continue beyond the hiring and job start of an employee. The HR department should maintain a good and continual line of communication with the employee and the supervisor to make sure the employee is performing well.

Fitter Snacker, like most companies, issues performance evaluations to new and current employees. The supervisor performs an initial evaluation and reviews it with the employee. After the review, the supervisor may modify the evaluation, which both the supervisor and the employee sign. The employee may submit a written response to the review, listing any disagreements or explanations. Other senior employees, such as the plant manager, may add a separate written comment, and should also review the performance evaluation and employee response. The complete package is then forwarded to the HR department, where all documents become part of the employee's file. These files are critically important when an employee consistently fails to perform adequately. If an employee must be terminated, the company needs sufficient documentation to demonstrate that the termination is warranted; otherwise, if the employee sues the company for wrongful termination, the company may have problems substantiating the termination decision. Because Fitter Snacker does not have an effective information system, it is frequently difficult to manage all of the performance evaluation data. This makes it difficult for the Human Resources Department to identify problems with an employee and take corrective action (such as counseling or a transfer) before the problems lead to termination of the employee. With Fitter Snacker's paper-based system, an employee's file can be viewed by only one person at a time, and it is possible to lose track of an employee's file—temporarily or permanently. Also, it is difficult to maintain proper control of sensitive personal information when it's maintained in paper files.

Figure 6-1 shows an employee data screen in SAP.

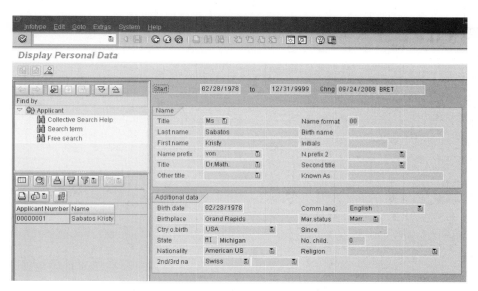

FIGURE 6-1   Personal data stored in SAP HR software

Employee turnover can be a significant problem. In its 2002 Cost-per-Hire (CpH) Staffing Metrics Survey, the Society for Human Resource Management reported that hiring costs for an employee may be as high as $70,000. This figure represents both the direct costs of hiring an employee and the less tangible losses that occur during a new employee's first year or so. While new employees are learning their jobs, other employees have to take time from their normal jobs to train them.

Another cost that is difficult to quantify is an employee's historical knowledge of the job, which is lost when he or she leaves a company. For example, if a purchasing manager leaves a company, then all of the manager's knowledge about supplier relations is lost. The company does have a record of the contract signed with the supplier, but details of the negotiations that led to the contract may not be documented. Such details can be crucial in successfully negotiating the next contract. The manager may have developed good relations with the supplier and know whom to contact when there are problems. These relationships are not specified as part of the purchasing manager position, but accrue over time with the individual holding the position. When companies experience high rates of turnover, they lose knowledge and skills that may be crucial to keeping them competitive.

Employee turnover is strongly related to job satisfaction and compensation. If employees have satisfying jobs and are well compensated, they are less likely to leave the company. Human Resources can help maintain a satisfying work environment through a number of means, such as holding training programs for supervisors and managers, conducting periodic employee satisfaction surveys, and gathering data from employee exit surveys. Human Resources also has a critical role to play in compensation, which should be related to the skills and tasks required by the job and the performance of the employee. An important function of the HR department is to make sure compensation levels are competitive and are applied fairly to all employees. Failure to do so can result in high rates of turnover as well as discrimination lawsuits.

ANOTHER LOOK

Discrimination Lawsuits

In February 2007, a federal appeals court approved class-action status for a discrimination lawsuit brought by seven women against Wal-Mart, claiming they were discriminated against in pay and promotion. It has been estimated that 1.6 million women who have worked for the large retailer since 1998 could join the lawsuit, which would make them the largest group ever to sue a company for discrimination. These women are claiming that they were denied promotion because of their gender, and that some were subjected to sexual harassment. In this class-action case, the group of seven women is bringing the suit against Wal-Mart on behalf of the larger group.

Wal-Mart is known for its close attention to data capture and storage. In this instance, these detailed data are being used against the company. Statistically, the case claims, Wal-Mart has been paying men more than women and has been promoting more men than women. Wal-Mart says its statistics do not support this claim. With its federal appeals court approval, this lawsuit has gone further than most. The U.S. Equal Employment Opportunity Commission reported that it had resolved 27,146 sex discrimination claims in 2003. Fifty-seven percent of the claims were dropped because they were found to have no reasonable cause, and only 10.6 percent resulted in settlements. Out of the original 27,146 claims, only 393 lawsuits were actually filed. At the time of the writing of this book, the Wal-Mart case had yet to be settled.

continued

The Boeing Corporation has also had a problem with pay discrimination. In 1996, the Labor Department's Office of Federal Contract Compliance Programs (OFCCP) ran a routine investigation of Boeing's Philadelphia plant. Because of Boeing's work on federal government contracts, the OFCCP had the right to audit Boeing's level of compliance with antidiscrimination laws. It did this by comparing the median pay and median job experience of male and female employees. Using this median analysis, the OFCCP report stated that Boeing demonstrated "a prima facie case of systemic discrimination concerning compensation of females and minorities." Boeing's response was to conduct its own analysis. Boeing initiated the Diversity Salary Analysis (DSA) project to develop a legally defensible statistical analysis of Boeing's pay practices, to counter the OFCCP's median analysis. Unfortunately for Boeing, the DSA project concluded in 1997 that "gender differences in starting salaries generally continue and often increase as a result of salary planning decisions." Boeing's own analysis showed that there was a pay gap for entry-level managers of $3,741.04.

In 2000, thirty-eight women filed a class-action lawsuit against Boeing, charging pay discrimination. Boeing's own salary studies supported the charges in the lawsuit, but Boeing claimed that the studies had been prepared at the direction of Boeing's lawyers and thus were protected by attorney-client privilege. While lawyers routinely prepare statistical studies to help defend a client against a lawsuit, the data contained in salary studies for business-related purposes (data that predate litigation and that are prepared by nonlegal executives) do not fall under the confidentiality protection of attorney-client privilege. On October 25, 2000, Judge George Pinkie ordered Boeing to release the salary studies, explaining that, "Legal departments are not citadels in which public business . . . may be placed to defeat discovery."

On May 17, 2004—two days before the discrimination case was scheduled to go to trial—Boeing made a settlement offer. After negotiation, in June 2004, Boeing agreed to pay $72.5 million to settle the case.

Questions:

1. Boeing settled its discrimination lawsuit for $72.5 million. What other costs were incurred by Boeing in association with this lawsuit? How do these costs compare to the financial cost of the settlement?

2. How could an integrated Human Resources information system be used to detect potential pay discrimination before it becomes systematic?

3. How would you design a compensation system so that pay discrimination is not likely to occur?

HUMAN RESOURCES WITH ERP SOFTWARE

Now that you are familiar with the numerous business processes required to manage a company's human capital, you can begin thinking about how ERP software can improve those processes, leading to overall improvements in a company's performance. With an integrated system, a company can store employee information electronically, eliminating the piles of papers and files that make the retrieval of information difficult or tedious. A good information system allows all relevant information for an employee to be retrieved in a matter of seconds. An integrated information system is a key component in this process.

ANOTHER LOOK

Major League Baseball Pioneers Sophisticated Data Analysis

Enterprise Resource Planning systems hold a wealth of data on a company—data that can be manipulated and analyzed using statistics. A wealth of data also exists on the sport of baseball, and those data are manipulated by statistics, too. Human resource managers can take a lesson from baseball when assessing employees' performance and hiring new employees. The Society for American Baseball Research uses statistical techniques called sabermetrics, which can analyze players on performance attributes other than the traditional batting averages. Measures are collected on the probability that a player will get a hit with other players on base—for example, by determining the number of times the player has done this in the past, and what type of hit he produced. Teams such as the 2007 World Champion Boston Red Sox are also using such measures to analyze their current roster and identify weaknesses, so they know which new players to recruit.

The sabermetrics method uses raw data, such as high-school and college records, family backgrounds, psychological profiles, and medical histories, all culled from a wide range of sources. Teams can also measure 11 attributes, including drive, endurance, leadership, self-confidence, emotional control, mental toughness, coachability, and trust. Using software, baseball scouts can look for players with the potential for good fielding, hitting, or even "hustle." Prospects are rated on physical qualities and baseball abilities such as running speed, hitting, fielding, and strength.

Companies such as Target are beginning to use this analytical approach to evaluate job applicants. Applicants complete in-store and online job applications that include questions to determine suitability for the position, as well as true-or-false statements such as "I would rather sit around and read a book than go to a party with lots of people," and "I don't act polite when I don't want to." Once the candidate has filled out the initial application, the system prompts the hiring manager with additional questions that probe more deeply into areas of concern.

Dow Chemical used to hire MBAs from Ivy League schools—until the company realized that these candidates demanded high salaries and few of them wanted to move to Midland, Michigan. Dow delved into the vast quantities of data in its PeopleSoft ERP system and found that its best candidates came from schools such as Michigan State, Brigham Young, and Purdue. Now Dow focuses its recruiting efforts at the institutions where it will be more successful. The cost to hire a new graduate, which exceeds $70,000, is now well spent.

Questions:

1. Assume you are the HR manager at Fitter Snacker. What metrics or measurements would you develop to assess the potential of an applicant for a production supervisor position? A sales manager position? A chief accountant?

2. How successful have the baseball leagues been in using sabermetrics? Research your answer on the Internet. Can you estimate how successful Fitter Snacker would be if it invested in a similar analysis?

Successfully using a Human Resources ERP system requires managing a significant amount of detailed information. The SAP ERP Human Resources (HR) module provides tools for managing an organization's roles and responsibilities, definitions, personal employee information, and tasks related to time management, payroll, travel management, and employee training. Advanced HR features of SAP ERP are discussed later in the chapter.

Most companies have an organizational chart or plan that helps define management responsibilities. Without an ERP system, the organizational chart defines only the managerial relationships among employees. With an ERP system, the organizational chart provides a structure with more detail than a typical organizational chart and supports HR tasks such as recruiting employees and planning organizational changes.

SAP ERP provides an Organization and Staffing Plan tool that is used to define a company's management structure and the positions within the organizational structure as a whole. The Organization and Staffing Plan tool also names the person who holds each position. Figure 6-2 shows how the Fitter Snacker organizational structure could be defined in SAP ERP. The figure shows that the Fitter Snacker organization consists of three main organizational units: Manufacturing, Marketing, and Administrative. The organizational units Accounting and Human Capital Management are part of the Administrative organization. Within the Human Capital Management organization are three positions: the HCM Manager and two Analysts.

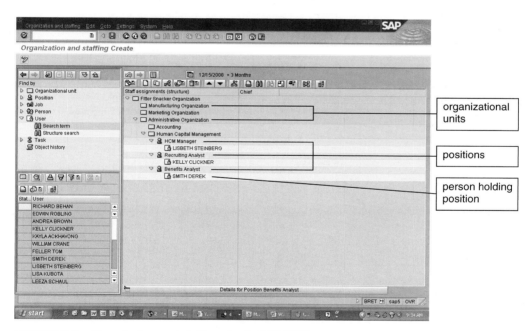

FIGURE 6-2   Organization and staffing plan in SAP ERP

SAP ERP distinguishes between **task**, **job**, **position**, and **person**. In SAP, an employee is a person who performs tasks, which can be assigned either to a job, which is a generic description of an employee's work responsibilities, or to the specific position that the individual person holds. Figure 6-3 shows the relationships among tasks, jobs, positions, and persons in a Marketing organization. The job of administrative assistant is assigned a number of tasks such as reviewing employee time charges, reviewing employee expense reports, and preparing monthly budget reports for the department. These are tasks that the company requires of any administrative assistant, whether that job is in Marketing, Engineering, Production, or another department. The job of administrative assistant can be defined once in SAP by assigning it tasks; then, that definition can be used to create administrative assistant positions in different organizational units. The administrative assistant job in Marketing is one position, while the administrative assistant job in Accounting is a different position. Additional tasks can be added to an administrative assistant position to tailor it to the specific requirements of the organizational unit. For example, in Figure 6-3, the position of Marketing Administrative Assistant has the marketing-specific task of preparing sales reports; an administrative assistant in Procurement would not be required to perform that task. In ERP, a person is the unique individual who fills a position.

FIGURE 6-3   Relationships among positions, jobs, tasks, and persons who fill positions

Figure 6-4 shows the screen in SAP ERP where tasks are assigned to jobs—in this case, the task Prepare Budget Report is assigned to the job Administrative Assistant. If the tasks associated with jobs and positions are well defined and current, it is easier for a recruiter to determine whether candidates have the qualifications for a job. Determining appropriate compensation for a position is also simplified if the tasks required for each position in a company are clearly and consistently defined.

Complete and accurate human resource data simplify a manager's duties. The Manager's Desktop tool within the SAP HR module provides access to all the Human Resources data and transactions in one location. Figure 6-5 shows the Personal Data portion of the Manager's Desktop. This area provides all of the data maintained in the Human Resources module for all employees who report to the manager. Human Resources data are very sensitive because they are related to employees' personal information, so controlling access

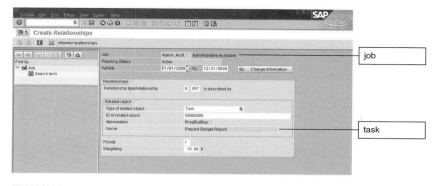

FIGURE 6-4 Assignment of a task to a job in SAP ERP

to them is critical. An advantage of an integrated information system over a paper-based system is that controlling access to data is automated; managers can use the system to determine which users should have access to various data.

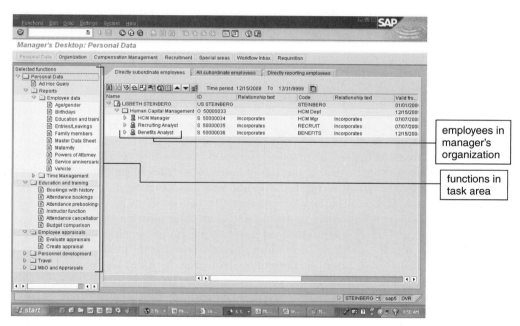

FIGURE 6-5 Manager's Desktop provides single-point access to HR functions

ADVANCED SAP ERP HUMAN RESOURCES FEATURES

Discussing in detail the many processes related to Human Resources is beyond the scope of this text; however, some of the advanced features of the Human Resources module in the SAP ERP system, including time management, payroll processing, travel and training coordination, are discussed below.

Time Management

Hourly employees, who are paid for each hour worked, must record the time that they work so that they can be paid. Salaried employees are not paid based on hours worked, but their time must usually be tracked as well. For cost-accounting purposes, it is important to be able to attribute an employee's time to a cost object (such as cost center, project, or production order), and any time not worked must be attributed to vacation or leave. The SAP ERP system uses Cross Application Time Sheets (CATS) to record employee working times and provide the data to applications that include:

- The SAP Controlling module, for cost management
- The SAP Payroll module, for calculating employee pay and transferring the data to the Financial Accounting module
- The SAP Production Planning module, to determine whether enough labor is available to support production plans

Payroll

Payroll is probably the most important Human Resources function. Employees are, not surprisingly, very particular about being paid the correct amount at the correct time! Many people live paycheck-to-paycheck, meaning that getting paid correctly and on time is crucial. Without proper management of the payroll process, employees might not be paid for all of the hours they worked, they might not be paid at the appropriate rate, or they might have too much or too little money withheld from their pay for taxes and benefits. Mistakes in payroll can cause significant job dissatisfaction. The two key processes in determining the pay an employee receives are calculation of the remuneration elements and determination of statutory and voluntary deductions. The **remuneration elements** of an employee's pay include the base pay, bonuses, gratuities, overtime, sick pay, and vacation allowances that the employee has earned during the pay period. The **statutory and voluntary deductions** include taxes (federal, state, local, Social Security, and Medicare), company loans, and benefit contributions. Properly determining the pay for an employee requires accurate input data and correct evaluation of remuneration elements and deductions.

The process of determining each employee's pay is called a **payroll run**. In the payroll run, the SAP ERP system evaluates the input data and notes any discrepancies in an **error log**. Payroll employees review the error log, make the necessary corrections, and repeat the payroll run until no errors are recorded. Then the payroll run is used to generate information for accounting, electronic funds transfers, employee pay statements, withholding tax payments, and other calculations.

Travel Management

Companies can spend a significant amount of money on employee travel, and managing travel and its associated expenses can be a significant task. A travel request, which may originate with the employee or the employee's manager, is the first step in the travel management process. Travel requests usually require management approval, and the level at which travel must be approved may depend on the duration and location of the travel. Once management has approved the travel, travel reservations must be made. Because airfare, hotel, and rental car costs can vary widely, companies frequently require employees to make reservations through either a company travel office or a travel agency under contract to the company. The employee must keep receipts for expenses incurred during the trip in order to complete an expense report and receive reimbursement. The SAP ERP Travel Management system facilitates this process by maintaining travel data for each employee, including flight, hotel, and car preferences, and integrating this data with the Payroll module (for reimbursements) and with the Financial Accounting and Controlling modules, to properly record travel expenses. Submitting an expense report can be simplified using a Web-based application that allows employees to submit reports through a Web browser.

ANOTHER LOOK

FXIS: From SAP Customer to Solution Provider

Photocopier manufacturer Fuji Xerox established Fuji Xerox Information Services (FXIS) in 1984 to create software for its computer-related products. Since that time, FXIS has taken over responsibility for managing Fuji Xerox's information systems and networks, and has expanded its business to include sales of computer and network equipment, as well as education and training in computer software applications.

In 1998, FXIS sought to improve its own internal software systems with the goal of being able to close its books (reconcile all of its accounting records) two days after the end of the month. To accomplish this task, FXIS chose to implement SAP ERP.

When planning its SAP implementation, FXIS decided that to meet its goal of closing its books in two working days it should use SAP for all of its business applications and get rid of legacy systems. Unlike some companies using SAP, FXIS emphasizes the integration of HR data. FXIS implemented the Sales & Distribution, Materials Management, Financial Accounting, Controlling, Project System, and Human Resources modules of the ERP system. To meet the two-day closing requirement, FXIS needed to have expense account information from its sales force and time charge information from its software developers and system administrators. To simplify the collection of travel and time charge data, FXIS created its Direct Input (DI) system, which makes it easy for users to enter their time charge data into the system through a Web interface. The DI system stores these data in its own database, which allows managers to review and approve the data. When the data have been approved, the information is transferred to the SAP ERP system.

continued

The DI system provides the advantage of a familiar browser interface for users who may not be comfortable using SAP ERP directly. It also does not require additional software or hardware. The DI system has proved so successful that FXIS has formed an ERP solution business and has developed standard templates for its DI system, which it is marketing to other SAP customers in Asia.

Questions:

1. What are the advantages and disadvantages of the FXIS decision to use SAP for all business processes and eliminate its legacy system?
2. How does the use of the DI system contribute to FXIS's goal of closing its books in two days? Do you think that FXIS experiences any problems because the data collected by the DI system are not available in real time?

Training and Development

The Personnel Development component of the SAP ERP Human Resources module supports the planning and implementation of employee development and training activities. Such education maximizes an employee's ability to contribute to the organization. Because advances in technology quickly render an employee's knowledge obsolete, employees will not be productive without continuing development and training efforts. In addition, many positions require certifications that must be updated, and continuing education is frequently required for recertification. Without an effective Human Resources information system, managing the training, development, and certification needs for a company's employees can be both time-consuming and error-prone.

In the SAP ERP system, employee development is driven by qualifications and requirements. **Requirements** are skills or abilities associated with a position, while **qualifications** are skills or abilities associated with a specific employee. Requirements and qualifications refer to the same concept from a different perspective. Using the Personnel Development tool allows a manager to compare an employee's qualifications with the requirements for a position to which the employee aspires. This comparison enables the manager to identify gaps and to plan development and training efforts to close the gaps. It can also serve as a basis for employee evaluation, and can motivate the employee by providing a goal and the means to achieve it.

One of the most important reasons for managing the development and training of employees is **succession planning**. A succession plan outlines the strategy for replacing key employees when they leave the company. The success of a company depends in large part on the skills, abilities, and experience of its management team. This is especially true for a small company like Fitter Snacker. Savvy customers have been known to avoid establishing long-term relationships with companies that do not have well-developed succession plans. The Career and Succession Planning components of the SAP ERP Human Resources module allow HR professionals to create, implement, and evaluate succession planning scenarios. HR departments use Career Planning with individual employees, identifying potential career goals and drawing up career plans. Companies use Succession Planning to find people to fill unoccupied positions. Succession Planning allows the human resources function to meet staffing requirements by identifying candidate employees within the company and ensuring that their training and development plans will prepare them for the new position when it becomes available. Using the career and succession planning tools in an ERP system ensures that HR will have accurate and timely employee and position data when developing the plans. The system allows changes in human resources (such as hiring new employees, current employees leaving, promotions) to be more easily integrated and tracked, and gives all important users easy access to the information.

ANOTHER LOOK

Management Succession Planning

Finding a chief executive officer (CEO) can be a tremendous challenge for a company. For many companies—including Coca-Cola, Xerox, and Procter & Gamble—finding a new CEO has been a process marked by long searches, poor choices, bad luck, and fumbled transitions. In other cases, succession planning has worked to ensure a smooth transition. When McDonald's Corporation CEO James Cantalupo died of a heart attack on April 19, 2004, succession planning allowed the board of directors to name Charles H. Bell as the new CEO within hours. Unfortunately, McDonald's succession planning capabilities were quickly tested a second time when Bell was diagnosed with cancer only one month after being named CEO. Bell was replaced by Jim Skinner in November 2004. Bell died on January 17, 2005.

At Juniper Networks, succession planning was called into action when vice president and chief information officer Alan Boehme had a car accident that left him with serious injuries. From the accident site, Boehme used his BlackBerry to send an e-mail message to Danny Moquin, then the vice president of operations and infrastructure, instructing him to take over Boehme's duties. Juniper had a modest succession plan on a spreadsheet—it only included the senior management. The plan was modest because the company was still in the midst of restructuring when the accident occurred.

A 2006 report on succession planning by Aberdeen Research found that 62 percent of companies have succession plans on paper or on a spreadsheet. This level of succession planning is insufficient. Succession planning goes hand-in-hand with disaster planning, but companies find disaster planning more important, even though accidents occur more frequently than hurricanes, earthquakes, or terrorist attacks—and people leave or get fired all the time. No one should be irreplaceable, yet most companies, like Jupiter, have only modest succession plans. Aberdeen found that 82 percent of companies have succession plans for executives, while 17 percent of companies have succession plans for their lower-level staff.

continued

ERP HR software can aid greatly in succession planning. At Juniper, Boehme has returned and is heading up the installation of Oracle's PeopleSoft for succession planning as part of its HCM software installation. Far more comprehensive than a simple list of names, HCM software incorporates information such as skill sets, experience levels, and work histories. Companies that don't have this information available often have minimal succession planning in place, or find succession planning to be unnecessarily challenging. Experts advise that companies should create a succession plan that incorporates every level of the organization. They also say that employees should be encouraged to take over for others who are on vacation, to gain experience in doing other jobs. And all employees should have their skill sets analyzed and recorded.

At Juniper, Moquin did a good job replacing Boehme, but the transition derailed momentum related to Juniper's growth and restructuring. Now, Boehme has 45 people assigned to the PeopleSoft implementation and is hoping to organize a more comprehensive succession plan.

Question:

1. What are the reasons that companies fail to create comprehensive succession plans?

ADDITIONAL HUMAN RESOURCES FEATURES OF SAP ERP

SAP ERP has to keep pace with rapidly changing social and legislative developments in the corporate world. The HR module has been expanded to include features that assist managers with HR tasks that have only recently become important to corporations.

Mobile Time Management

Many employees, especially sales personnel who spend a significant amount of time on the road, may not have regular access to a PC. Mobile Time Management allows employees to use cellular phones to record their working times, record absences, enter a leave request, and check their time charge data.

Management of Family and Medical Leave

The Human Resources module reduces the administrative burden imposed by the federal Family and Medical Leave Act (FMLA) of 1993. The HR system can now determine whether an employee is eligible to take FMLA absences and automatically deducts those absences from the days the employee takes from allowable leave.

Domestic Partner Handling

Many companies provide benefits for domestic (unmarried) partners. The Human Resources module now supports the management of benefits for domestic partners and their children. The system now provides more flexibility in customizing dependent coverage options for health plans, eligibility for enrollment of dependents, and designation of beneficiaries.

Administration of Long-Term Incentives

An outgrowth of the Sarbanes-Oxley Act (see Chapter 5) is that companies must account for the expected costs that occur as a result of long-term incentives such as the exercising of stock options. The Human Resources module now provides more options for processing long-term incentives. Integration with the SAP Payroll module enables companies to calculate taxes accurately when employees exercise incentives and sell their shares in the company. SAP can share the incentive data with Accounting so that Accounting can do the necessary reporting.

Personnel Cost Planning

Changes in an organization (including expansions, acquisitions, and downsizing) can have an impact on employee-related expenses, which are usually a significant portion of a company's costs. The Personnel Cost Planning tool allows HR personnel to define and evaluate planning scenarios to generate cost estimates. Performing cost planning and simulation allows HR to forecast cost estimates by integrating data with other SAP ERP modules.

Management and Payroll for Global Employees

The management of global employees involves many complicated issues, including relocation plans, visas and work permits, housing, taxes, and bonus pay. SAP ERP has enhanced features to support the management of these issues, with customized functionality for over 50 countries, allowing payroll processes to meet current legal regulations and collective bargaining agreements in the local business environments.

Management by Objectives

The concept of management by objectives (MBO) was first outlined by Peter Drucker in his 1954 book *The Practice of Management*. In MBO, managers are encouraged to focus on results, not activities, and to "negotiate a contract of goals" with their subordinates without dictating the exact methods for achieving them. SAP ERP provides a comprehensive process to support the MBO approach that incorporates performance appraisal. The appraisal results can affect an employee's compensation, generating annual pay raises that can be either significant or insignificant, depending on the employee's performance. The MBO process in SAP ERP also allows managers to include the results of achieved objectives in the employee's qualifications profile.

Chapter Summary

- Employees are among a company's most important assets. Without qualified and motivated employees, a company cannot succeed.

- The Human Resources department has the primary responsibility of ensuring that the company can find, evaluate, hire, develop, evaluate, and compensate the right employees to achieve the company's goals. HR is also responsible for employee training and development, succession planning, and termination.

- Managing, sharing, controlling, and evaluating the data required to manage a company's human capital are simplified by an integrated information system.

- Additional features of the SAP HR system address today's changing technology and legislation.

Key Terms

Error log	Remuneration elements
Human capital management (HCM)	Requirements
Job	Short list
Payroll run	Statutory and voluntary deductions
Person	Succession planning
Position	Task
Qualifications	

Exercises

1. Describe a position in a company that you would like to have. What type of information must be collected to determine if a candidate is appropriate for this job? List the skills that you think would be required for this position. Suppose you are designing a system to summarize information from resumes submitted to a company's Human Resources department. Create a list of the information that you think would be useful to collect from the resumes.

2. Describe a position in a company that you would like to have after five years of work experience. List the requirements that you think would be necessary to hold this position. List the qualifications that you currently possess. Describe how you plan to obtain the qualifications necessary to hold the position.

3. Suppose you are a manager of Fitter Snacker's Sales department. What Human Resources information do you think you would need to manage your sales force?

4. List the steps in a typical recruiting process. Highlight the steps that involve interaction with the potential job candidate. Identify problems in the process that might lead a candidate to develop a negative opinion of the company. How might an effective information system reduce the potential for these problems? Incorporate into your answer experiences you may have had in looking for a job.

For Further Study and Research

Ackman, Dan. "Wal-Mart and Sex Discrimination By the Numbers." Forbes.com, June 23, 2004. http://www.forbes.com/careers/2004/06/23/cx_da_0623topnews.html.

CNNMoney.com. "Wal-Mart to appeal discrimination suit status." CNNMoney.com, February 6, 2007. http://money.cnn.com/2007/02/06/news/companies/walmart/index.htm.

Duvall, Mel. "Boston Red Sox: Backstop Your Business." *Baseline*, May 14, 2004. http://www.baselinemag.com/article2/0,1397,1590697,00.asp.

Hines, Matt. "Postal Service seals big SAP deal." CNET.com, August 23, 2004. http://news.com. com/Postal+Service+seals+big+SAP+deal/2100-1012_3-5319934.html?tag=item.

Holmes, Stanley, and Mike France. "Coverup At Boeing? Internal documents suggest a campaign to suppress evidence in a pay-bias lawsuit." *Business Week,* June 28, 2004.

Lavelle, Louis. "How to Groom the Next Boss." *Business Week,* May 10, 2004.

Lynch, C. G. "Smash-Up: How a Violent Car Crash Provided Lessons in Business Continuity and Succession Planning." *CIO*, July 11, 2007.

Singer, Michael. "SAP Delivers for the Mailman." Internetnews.com, August 25, 2004. http://www. internetnews.com/ent-news/article.php/3399361.

Taleo.com. "iLogos Study Shows Majority of Global 500 Companies Recruit on the Careers Web Site." Taleo.com, April 8, 2002. http://www.taleo.com/news/press/ilogos-study-shows-majority-global-500-85.html.

"US Postal Service Selects mySAP Business Suite." *SAP & Partner News,* August 28, 2004.

Wooldridge, Adrian. "A Survey of Talent." *The Economist*, October 7, 2006.

CHAPTER **7**

PROCESS MODELING, PROCESS IMPROVEMENT, AND ERP IMPLEMENTATION

LEARNING OBJECTIVES

After completing this chapter, you will be able to:

- Use basic flowcharting techniques to map a business process.
- Develop an event process chain (EPC) diagram of a basic business process.
- Evaluate the value added by each step in a business process.
- Develop process improvement suggestions.
- Discuss the key issues in managing an ERP implementation project.
- Describe some of the key tools used in managing an ERP implementation project.

INTRODUCTION

The underlying theme of this text is the management of business processes. In this chapter, we explore tools, such as flowcharts and even process chains, that can be used to describe processes. Next, you learn about how these tools, which are not specific to ERP, can help managers identify process elements that can be improved. We finish by describing the role these process-modeling tools play in ERP implementation projects.

PROCESS MODELING

By now, it should be clear that business processes can be quite complex. Individuals with various skills and abilities are responsible for executing business processes. In order for business processes to be effective (achieve the desired results) and efficient (achieve the desired results with the minimum use of resources), they must be clearly defined, and individuals must be adequately trained to perform their roles and to understand how their roles fit within the business process.

We have used a range of terms, including flowchart, to describe methods of representing processes. We will use the term **process model** to describe any abstract representation of a process. A process model can be as simple as a diagram with boxes and arrows or as complex as computer software that allows for simulation of the process. Process-modeling tools provide a way to describe a business process so that all participants can understand the process. Frequently, process models are developed by a team of employees involved in the process. The interaction required to develop a process model often reveals misunderstandings and ensures that all team members are "on the same page." Graphical representations are usually easier to understand than written descriptions. A well-developed process model provides a good starting point for analyzing a process so that participants can design and implement improvements. Process models also document the business process, making it easier to train employees to support the business process.

Flowcharting Process Models

Flowcharts are the simplest of the process-modeling tools. A **flowchart** can be defined as any graphical representation of the movement or flow of concrete or abstract items—materials, documents, logic, and so on. Flowcharts originated with computer programmers and mathematicians, who used them to trace the logical path of an algorithm. In the early days of computer programming, computer resources were limited, and executing a program used considerable resources. As a result, most programmers spent a significant amount of their time clearly defining the logic of their programs, using flowcharts before actually writing the code and testing the program.

A flowchart is a clear, graphical representation of a process from beginning to end, regardless of whether that process is an algorithm or a manufacturing procedure. Flowcharting has been used since the 1960s in business applications to help businesspeople visualize workflows and functional responsibility within organizations. Today the term **process mapping** is used interchangeably with flowcharting, the distinction being that process mapping specifically refers to the activities occurring within an *existing* business process. Process mapping develops an "as is" representation of the process, with a goal of exposing weaknesses that need to be addressed. Once a company develops a process map, it can perform a **gap analysis**, which is an assessment of disparities between an organization's current situation and its long-term goals.

Flowcharting uses a standardized set of symbols to represent various business activities. Few symbols are required to define a business process. Figure 7-1 shows the basic flowcharting symbols. You can use a wide range of symbols for process mapping, but the basic set shown in Figure 7-1 is sufficient to describe even a complicated business process. Using a few simple symbols places the focus on the process, not on the tool used to represent it.

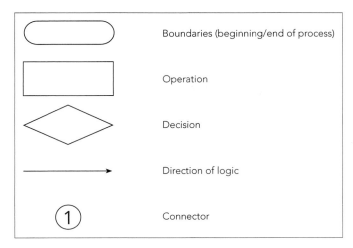

FIGURE 7-1 Basic flowcharting symbols

The following description of the process Fitter Snacker uses to reimburse sales personnel for their business expenses illustrates process mapping.

Fitter Snacker Expense Report Process

After a Fitter Snacker salesperson, Maria, incurs travel expenses on her credit card, she completes a paper expense report, makes a copy for her records, attaches receipts for any expenses over $25, and mails it to her zone manager at the branch office. The manager, Kevin, reviews it and either approves the report or mails it back to Maria with a note asking for an explanation, verification, or modification. Once Kevin approves the expense report, he mails it to the corporate office. After the administrative assistant sorts the mail at the corporate office, she forwards the expense report to the accounts payable (A/P) clerk, who performs a preliminary check of the report. The clerk contacts the zone manager for any necessary clarification, then forwards the expense report to the expense report auditor, who reviews it. If there is a problem with the report, the auditor mails it back to Maria, who revises and returns it. Then the auditor enters the report into Fitter Snacker's PC-based accounting system and files a hard copy with the receipts in a filing cabinet, organized by employee name.

At the end of each week, an A/P clerk uses the PC-based accounting system to print payroll checks, payments to suppliers, and expense reimbursement checks. When Maria receives her reimbursement check, she deposits it into her checking account and mails a payment to the credit card company, which credits her card account. Figure 7-2 shows the process map for the first part of the current Fitter Snacker expense-reporting process.

One of the most important decisions to make in process mapping is to define the process boundaries. The **process boundaries** define which activities are to be included in the process, and which are considered part of the environment—external to the process. It is important to clearly define the activities that are part of the process and those that are external to the process. Many activities related to sales employee expenses are considered outside of the boundary of the process map shown in Figure 7-2. For example, a

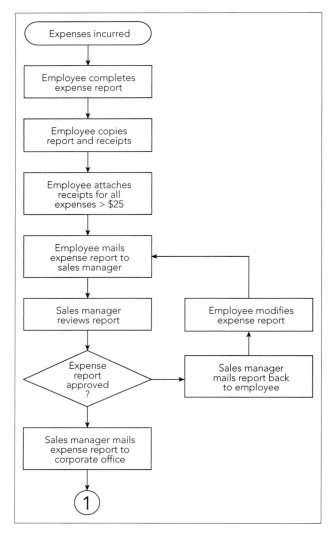

FIGURE 7-2   Partial process map for Fitter Snacker expense-reporting process

company might provide cash advances or issue corporate credit cards before employees incur travel expenses. Employees might need to make reservations for air travel or hotels through a corporate travel office. The company might want employees to use a preferred hotel, or might have specific policies regarding car rentals, including preferred rental companies, approved car classes, insurance, and prepaid gas. While all of these considerations are important and should be documented somewhere, the process mapped in Figure 7-2 is the *expense-reporting process*; the process boundaries do not include these additional travel-related activities. The process map begins after travel expenses have been incurred and ends with the receipt by the salesperson of a refund check. These starting and end points are the process boundaries for the expense-reporting process.

In Figure 7-2, the beginning process boundary is represented by the oval figure containing the text "Expenses incurred." All processes should have only one beginning point and one ending point. After the Expenses incurred oval, the process map shows four operation blocks that define the tasks performed by the employee in completing the expense report. The number of operation blocks and the level of detail in the descriptions are a matter of the user's preference and depend on the purpose for which the process map is created. If the process map will be used to improve a business process, and the members of the process improvement team are familiar with the process, then less detail is needed. In fact, too much detail could obscure the key features of the process. On the other hand, the process map might be used to document the process for training new sales employees. In that case, more detail is needed so that new employees can use the process map to follow the process properly.

Figure 7-2 also contains a decision diamond. A decision diamond asks a question that can be answered with "yes" or "no." In the figure, the decision diamond asks whether the sales manager approves the expense report. There are only two possible options—yes and no. It is tempting for the novice to create process maps with decision diamonds that have more than two outcomes. Doing so can lead to confused logic—all business processes can be defined using one or more decision diamonds, each with only two outcomes.

Finally, because Figure 7-2 only shows part of the expense-reporting process, the flowchart ends with a connector. Most business processes are too complicated to fit on a single sheet of paper. Connectors provide a way to continue process maps from one sheet to the next.

Exercise 7.1

Complete the process map for the Fitter Snacker expense report process started in Figure 7-2, using the process-mapping symbols shown in Figure 7-1.

Extensions of Process Mapping

The development of computer technology, specifically high-quality graphical interfaces, has allowed process-mapping tools to evolve beyond the simple symbols of flowcharting. One helpful tool is **hierarchical modeling**, which is the ability to flexibly describe a business process in greater or less detail, depending on the task at hand. Figure 7-3 illustrates a hierarchical model of Fitter Snacker's expense report process.

It is important to document (for training purposes) the detailed steps that the salesperson follows to complete the expense report; however, the details can make the process map cumbersome for process improvement activities. In hierarchical modeling, the four steps that the salesperson follows to complete the expense report can be condensed into one step, designated as Employee prepares expense report. Modeling software that supports hierarchical modeling provides the user with the flexibility to move easily from higher-level, less detailed views to the lower-level, more detailed views. Hierarchical modeling can aid in process mapping by allowing a user to create a broad, high-level view of a process and then add more detail as the process is analyzed.

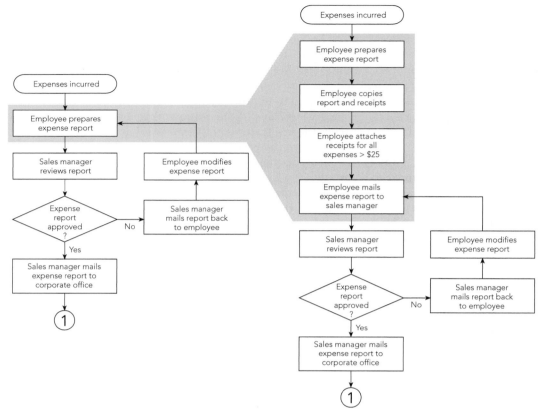

FIGURE 7-3 Hierarchical modeling

Another widely used and widely recognized type of process-mapping technique is **deployment flowcharting**. This type of flowchart depicts team members across the top, with each step aligned vertically under the appropriate employee or team. Figure 7-4 shows the Fitter Snacker expense report process as a deployment flowchart. This type of process map is also referred to as a **swimlane flowchart**. This process-mapping technique has the advantage of clearly identifying each person's tasks in the process.

Event Process Chain (EPC) Diagrams

ERP software such as SAP consists of business applications that support business processes. SAP's software supports hundreds of business processes, and SAP has developed graphical models for many of these business processes, using the **event process chain (EPC)** format. The EPC format uses only two symbols to represent a business process. The advantage of the EPC format is that it matches the logic and structure of SAP's ERP software design. The EPC modeling technique is available as a software tool through the IDS Scheer company as the ARIS (Architecture of Integrated Information System) toolset.

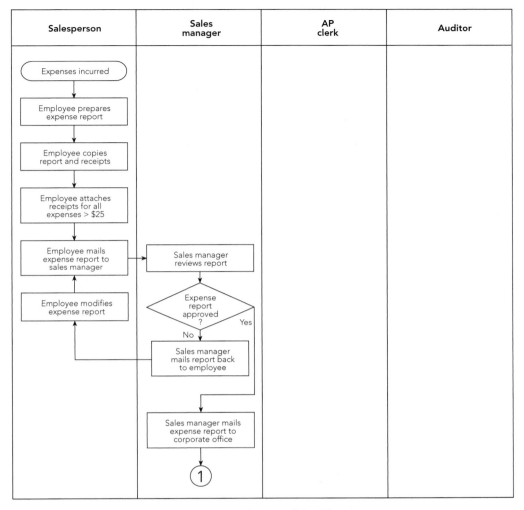

FIGURE 7-4 Deployment, or swimlane, flowcharting of the FS expense report process

The two structures used in EPC modeling to represent business processes are events and functions. As shown in Figure 7-5, events reflect a state or status in the process, and functions represent the part of the process where change occurs.

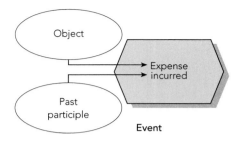

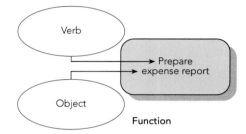

FIGURE 7-5　EPC components

Unlike flowcharting, EPC modeling enforces a strict structure. EPC software enforces an event-function-event structure. A standardized naming convention for functions and events is also used in EPC modeling. For example, the convention for naming events is Object → Past Participle:

Object	Past Participle
Expense	Incurred
Expense report	Approved
Hard copy	Filed

For functions, the naming convention is Verb → Object:

Verb	Object
Prepare	Expense report
Review	Expense report
Mail	Refund check

Figure 7-6 shows a simple EPC diagram for part of the Fitter Snacker expense report process. EPC diagrams begin and end with events. Furthermore, events must be followed by functions, and functions must be followed by events.

In addition to direct connection of events to functions, EPC diagrams employ three types of branching connectors. Branching occurs when logic either comes from more than one source or proceeds to more than one potential outcome. The three connector types are AND, OR, and Exclusive OR (XOR). Figure 7-7 shows an application of the OR connector, which indicates that after the payment is processed, the salesperson is notified, or the sales manager is notified, or both the salesperson and sales manager are notified.

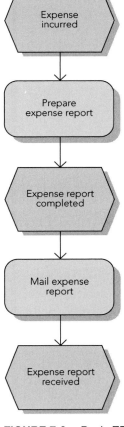

FIGURE 7-6   Basic EPC layout

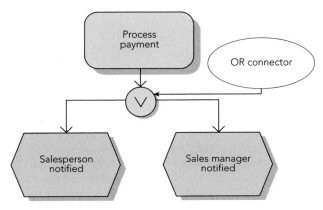

FIGURE 7-7   OR connector

Process Modeling, Process Improvement, and ERP Implementation

Figure 7-8 shows how the AND connector functions. The figure indicates that the expense report must be recorded and a hard copy must be filed.

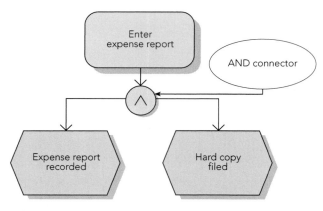

FIGURE 7-8   AND connector

Figure 7-9 shows how the XOR connector can be used to represent the manager's decision on whether to approve the expense report. The XOR connector is exclusive, so Figure 7-9 indicates that only one event can occur after reviewing the expense report: it is approved or it is not approved.

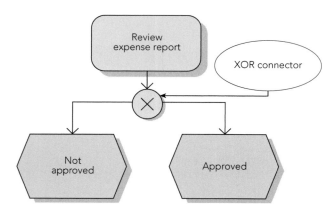

FIGURE 7-9   XOR connector

Figures 7-7 through 7-9 show one function connecting to two events using a branch connector, but it is not always the case that functions lead into the connector, nor do multiple events always follow the connector. For example, Fitter Snacker could require that salespeople complete expense reports at the end of a short sales trip, but at the end of each week if a trip lasts more than one week. This condition is illustrated in Figure 7-10, where the preparation of the expense report can be triggered by either of two events: the end of the trip *or* the end of the week. It is also possible that more than two events could trigger the function.

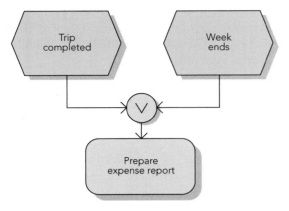

FIGURE 7-10   OR connector with two triggering events

Figure 7-11 shows all possible connection combinations. Note that it is not possible to have a single event connect to multiple functions with OR or XOR connectors, because events represent a status or state. Because OR and XOR connectors require a decision, they must be preceded by a function, so that a decision can be made.

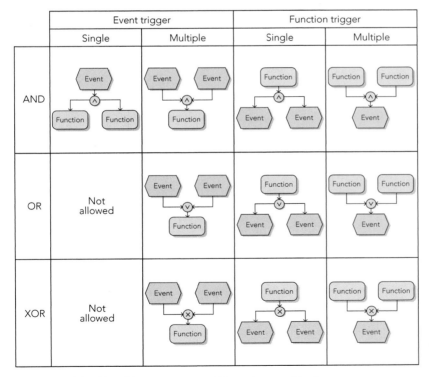

FIGURE 7-11   Possible connector and triggering combinations

Process Modeling, Process Improvement, and ERP Implementation

Finally, Figure 7-12 shows splitting and consolidating of a path through the process. In this case, the Fitter Snacker salesperson can submit her expense report online if she has Internet access; otherwise, she must send in a paper report. Note that the type of branch connector that splits the path also must be used to consolidate it.

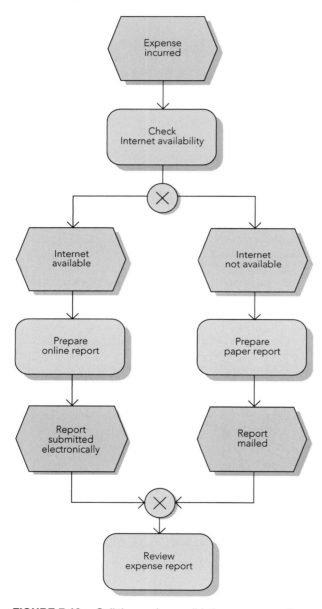

FIGURE 7-12   Splitting and consolidating process paths

The basic EPC diagram can be augmented with additional information. For example, Figure 7-13 shows the first part of an EPC diagram for the Fitter Snacker expense report process that also shows data elements (unapproved multicopy expense reports) and organizational elements (salesperson, sales manager). The additional elements allow for a more complete description of the process, documenting the "who" and "what" aspects of the process.

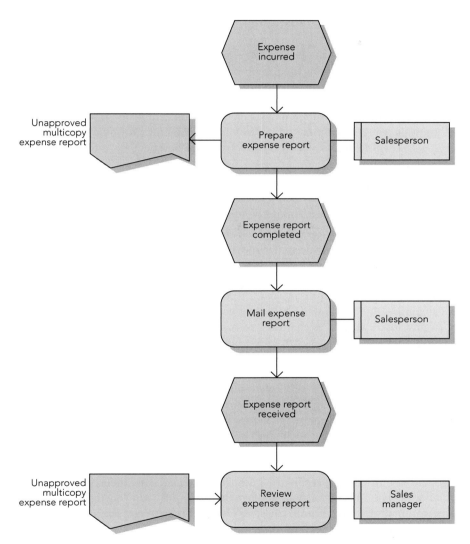

FIGURE 7-13 EPC diagram with organizational and data elements

Exercise 7.2

Using the above description, complete the EPC diagram of the Fitter Snacker expense report process shown in Figure 7-13. Add both the logic elements (events, functions, and connectors) and the organizational elements.

PROCESS IMPROVEMENT

Process-mapping tools provide the ability to describe business processes in a universally understood format. Generally, the task of mapping a process requires a team consisting of key personnel who are involved in the process. Frequently, the act of accurately describing the process, and understanding how the functional areas interact, make it obvious to the team what steps are necessary to improve the process. This is especially true for organizations that have focused on functional responsibilities, and not on business processes.

Using the simple technique of value analysis can also generate process improvement ideas. In **value analysis**, each activity in the process is analyzed for the value it adds to the product or service. The **value added** is determined from the perspective of the customer. Activities can add:

- *Real value*: Value for which the customer is willing to pay
- *Business value*: Value that helps the company run its business
- *No value*: An activity that should be eliminated

Activities that cost more than their value added should be improved. The Fitter Snacker expense report process does not provide real value, because Fitter Snacker's customers do not care whether sales employees receive prompt and accurate reimbursement of their business expenses. However, the expense report process does provide business value, and it should provide this value at a minimum cost. Evaluating the value of a business activity is not a hard science. Determining the value of a good or service is easy—it's what someone is willing to pay for it. Applying this idea to a part of a business process is more challenging, because parts of a process can't be "sold" on the open market. While a challenging task, evaluating each activity on the basis of value provided can highlight opportunities for improvement.

The value analysis concept can be expanded to an evaluation of both the time and cost of each process step. For each step in the current process, you would estimate the actual time and cost. Then you would estimate the value-added time and cost—determining how much of the actual time is adding value and how much of the cost is worth paying for.

We will use a Fitter Snacker process to illustrate value analysis. The company's *mail expense report* function could cost upwards of $50, including not just the cost of the envelope and postage, but also the time spent by the salesperson to mail the expense report. The value analysis includes elapsed time for mailing the expense report—the length of time from when the salesperson mails the report until the sales manager receives it. This elapsed time should include the time it takes the salesperson to find a mailbox, the time the postal service takes to deliver the expense report to the company headquarters, plus the time it takes the company's internal mail system to deliver the expense report to the sales manager.

Suppose that for Fitter Snacker the elapsed time is typically three days. To perform value analysis, you would determine how much of this time and cost is value-added. To determine this value, you must view the activity *mail expense report* in terms of what is actually being accomplished. It is not the physical transmission of paper that matters. Mailing the expense report is a means to transmit expense data and documentation. E-mail minimizes the time and cost required to transmit data; therefore, mailing an expense report at a cost of $50, to arrive in three days, is only providing value worth pennies (which is the cost to send an e-mail) in a function that it should be possible to execute within seconds. Looking at the current time and cost of each process step, then asking, "What is actually being accomplished and what is the value?" can help identify areas for process improvement.

Each step in a business process should be challenged to determine if it is providing value. H. James Harrington, in his book *Business Process Improvement*, suggests that companies ask the following questions about their business processes to identify areas for improvement:

- Are there unnecessary checks and balances?
- Does the activity inspect or approve someone else's work?
- Does it require more than one signature?
- Are multiple copies required?
- Are copies stored for no apparent reason?
- Are copies sent to people who do not need the information?
- Is there unnecessary written correspondence?
- Are there people or agencies involved that impede the effectiveness and efficiency of the process?
- Do existing organizational procedures regularly impede the efficient, effective, and timely performance of duties?
- Is someone approving something they already approved (for example, approving capital expenditures that were approved as part of a budget)?
- Is the same information being collected at more than one time or location?
- Are duplicate databases being maintained?

Harrington also suggests concepts that can be used to improve processes:

- Perform activities in parallel, for example, approvals.
- Change the sequence of activities.
- Reduce interruptions.
- Avoid duplication or fragmentation of tasks.
- Avoid complex flows and bottlenecks.
- Combine similar activities.
- Reduce the amount of handling.
- Eliminate unused data.
- Eliminate copies.

Evaluating Process Improvement

While identifying process improvements is challenging, implementing them is even more challenging. Disrupting the current process to make changes can be costly and time-consuming, and managers are frequently reluctant to risk trying process improvement ideas—especially if the ideas require significantly different ways of doing things. One way to overcome this risk is to use dynamic process modeling to evaluate process changes before they are implemented. **Dynamic process modeling** takes a basic process flowchart and puts it into motion, using computer simulation techniques to facilitate the evaluation of proposed process changes. Computer simulation uses repeated generation of random variables (such as customer orders) that interact with a logical model of the process to predict the performance of the actual system. These models can estimate the performance of the system, using measures such as cycle time (how long the process takes), productivity, total cost, idle time, and bottlenecks.

ANOTHER LOOK

Business Process Innovation at Nova Chemicals

Nova Chemicals has used business process innovation to move from a function-oriented company to a process-oriented company. According to John Wheeler, CIO of Nova Chemicals, "**Business process innovation (BPI)** is the process of improving processes. BPI is based on understanding the way you work. Once you understand the way you work, you can begin to improve the way you work." As part of its BPI initiative, Nova has used IDS Scheer's ARIS toolset to document its business processes. The ARIS toolset is a graphical process-mapping tool that helps companies define business processes and the relationships between the processes and the people who execute them. Nova has seen the ARIS toolset as a "huge enabler" in BPI. The tool requires structure and discipline to use, but allows companies to understand all of their processes, not just the workflow. Wheeler estimates that in the early days of computer information systems, technology was 75 to 80 percent of the cost of a project, while today technology is only 10 to 15 percent of the project's cost. Wheeler estimates that 30 to 40 percent of the budget for a business process improvement project is spent on understanding the current process.

Nova has seen considerable success using BPI in its sales process. In this project, Nova gathered top salespeople and had them document the sales process using the ARIS toolset. Nova discovered that every salesperson performed the process differently, and in facilitating innovation sessions, managers were able to develop the best practices for use throughout the company, taking the best ideas from each participant. Wheeler has observed that "thought leaders"—the people who contribute most to generating process improvement ideas—can come from anywhere in the organization, and it is frequently surprising who the thought leaders turn out to be.

Wheeler sees BPI as just the next step in the evolution of process improvement methods. Other methods for improving processes include quality circles, continuous improvement, and business process reengineering. In many ways, BPI resembles quality circles, the quality improvement technique pioneered in Japan in the 1970s. In **quality circles**, employees in a department have regular team meetings to discuss problems and collaboratively develop solutions. Quality circles were followed by the **continuous improvement** quality philosophy, again imported from Japan, which prescribed systematic and repeated improvement efforts. The concept of **business process reengineering** was popularized by Michael Hammer and James Champy in their 1993 book *Reengineering the Corporation*. Unlike continuous improvement, which stressed repeated improvement activities to achieve gradual improvements, business process reengineering recommends radical change to achieve radical improvements. BPI is the latest evolution of improvement methodologies.

continued

Some people might see the continually changing methods merely as fads that allow consultants to sell new consulting services. Wheeler doesn't see it that way. Innovations in methods and techniques of process improvement can keep innovation fresh and new. To Wheeler, BPI is itself a process, and all processes can be improved.

Questions:

1. What role does process mapping play in business process improvement? Is there an advantage to using a structured approach like IDS Scheer's ARIS toolset?
2. How would you form a team to improve a business process? How would you manage the project? What tools would you use?
3. Can you think of any ways in which technology can contribute to further improvements in the process of improving business processes?

ERP WORKFLOW TOOLS

Most business processes are performed regularly, enabling employees responsible for the process to become efficient in the tasks involved in the process. For example, the sales order process is fundamental to a manufacturing business; the salespeople, sales order clerks, warehouse managers, accounts receivable clerks, and others are spending most of their day supporting the process. If the process is efficiently designed and managed, and the employees are properly trained, workers will experience enough repetition to become efficient in their daily tasks.

Many business processes, however, are performed sporadically. The effectiveness of these processes can be poor, especially when the processes apply to more than one functional area. Many times, the work "falls through the cracks," not necessarily through negligence, but due to a lack of repetition. For example, the process of establishing credit limits occurs occasionally, and requires coordination between Sales, which identifies new customers and gathers basic data (contact names, addresses, terms and conditions) and Accounts Receivable, which must evaluate the customer's credit history to establish a credit limit. Unless employees manage the process of establishing a credit limit properly, a new customer's order may be blocked for an unacceptable length of time. For sporadic processes, a workflow tool can automate the process to ensure that the tasks are performed in a timely and correct manner.

Workflow tools are software programs that automate the execution of business processes and address all aspects of a process, including the process flow itself (the logical steps in the business process), the people involved (the organization), and the effects (the process information). ERP software provides a workflow management system that supports and speeds up business processes. It enables employees to carry out complex business processes and track the current status of a process at any time.

The SAP ERP workflow tool integrates organizational data (which indicate how an authorized worker is supposed to perform a transaction) with business transactions. SAP's internal e-mail system lets you use workflow to view links between work and various transactions. These links, called **workflow tasks**, can include basic information, notes, and

Process Modeling, Process Improvement, and ERP Implementation

documents, as well as direct links to business transactions. The SAP system can monitor workflow tasks, and if the tasks are not completed on time, the workflow system can automatically take various actions, including changing the workflow task priority and sending e-mail reminders to the employees responsible for the work.

Consider vacation requests. Employees performing critical operational tasks may be required to request time off from work in advance. This is a fairly irregular process that can create operational and/or employee morale problems if not handled properly. As such, it is a prime candidate for a workflow tool. In SAP, the Workflow Builder is used to define the process behind a workflow. In addition to defining the business process steps, the software identifies individuals involved in the process and sets other process parameters. Figure 7-14 shows a Workflow Builder screen that manages an employee's request for time off.

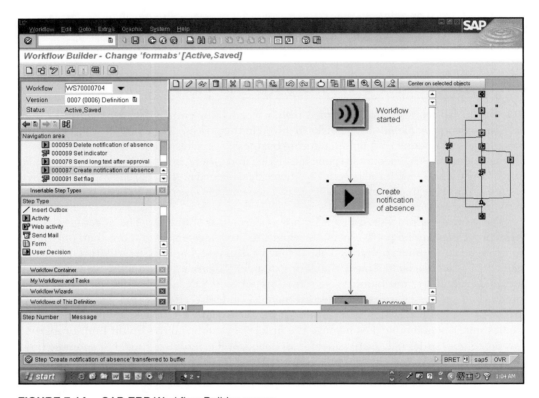

FIGURE 7-14   SAP ERP Workflow Builder screen

Figure 7-15 shows the first step in this process, the absence request screen that the employee uses to request time off from work. Once the employee completes the screen requesting time off, the request becomes a workflow item in the Business Workplace for the employee's supervisor. The Business Workplace is a collection of SAP programs that provides a number of functions, including e-mail, a calendar, and workflow, similar to other comprehensive communication/collaboration software packages like Microsoft Outlook or IBM's Lotus software.

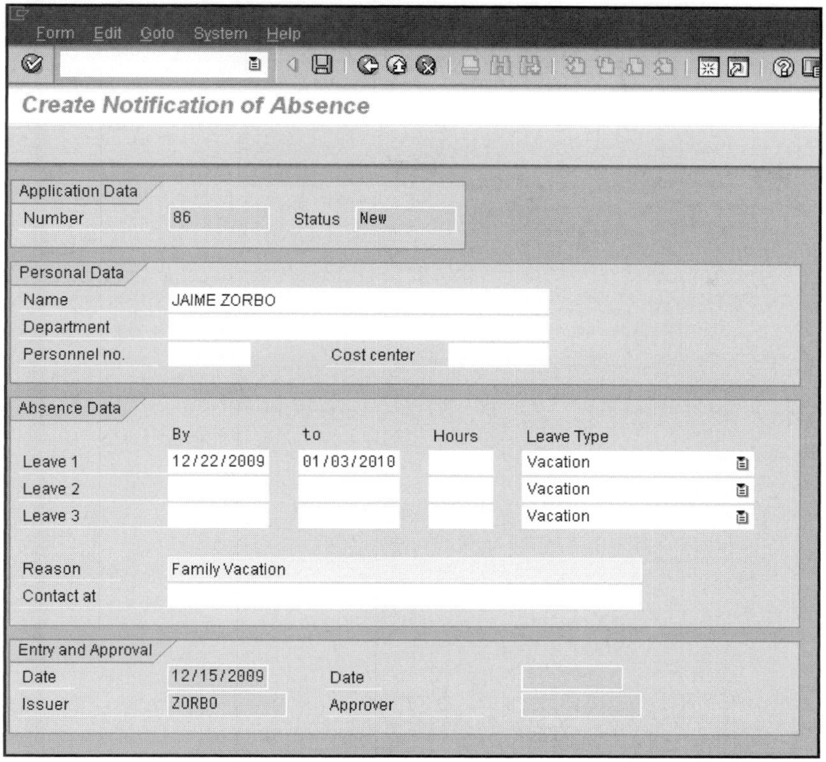

FIGURE 7-15 Absence request screen

Figure 7-16 shows a request for time off in the workflow inbox of the manager's Business Workplace. From this screen, the manager can either approve or deny the request. If the request is approved, the employee receives notification of the approval by e-mail. If the request is denied, the manager can include a note explaining the rationale. The rejected request will be sent to the employee's workflow inbox, where that person can either modify the request or cancel it.

Workflow provides a number of useful features. Employees can track the progress of workflow tasks, reviewing their status at any time. The system can be programmed to send reminders to the employee(s) responsible for a task after a predetermined period of time. Managers can build flexibility into workflow tasks regarding who can perform the task. For example, suppose that the Accounts Receivable (A/R) department has three employees who can set credit limits. You could create a workflow task to set a new customer's credit limit and send it to the workflow inbox of all three A/R employees. When one of the employees completes the task, the system removes it from all three workflow inboxes. You can also generate statistics on the number of workflow tasks handled by each employee and the average time taken to complete the tasks, resulting in better staffing decisions.

For regular, day-to-day business processes, workflow tools are not required. But for sporadic processes that are repeated frequently enough to justify the development costs, workflow tools are a powerful way to improve process efficiency and effectiveness.

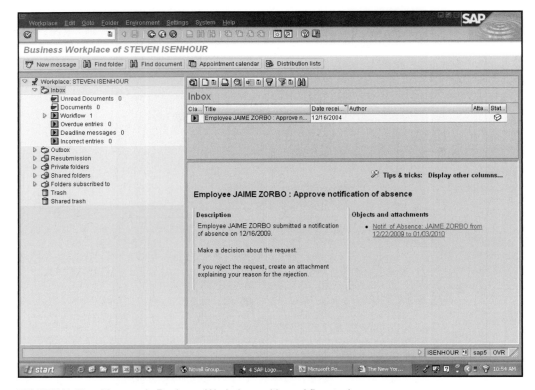

FIGURE 7-16   Manager's Business Workplace with workflow task

IMPLEMENTING ERP SYSTEMS

You learned about some of the issues and challenges in implementing ERP in Chapter 2. Implementation was particularly challenging in the late 1990s, as many firms rushed to implement ERP systems to avoid the Y2K problem. The explosive growth in demand for ERP implementations at that time caused a significant shortage of experienced consultants. Since 2000, the pace of implementations has slowed considerably. Most Fortune 500 firms have implemented an ERP system. The current growth area in ERP implementations is in the small to midsized business market, and vendors have been developing products, including Microsoft Great Plains and SAP Business One, tailored to this market.

Recall that the implementation of ERP is an ongoing process, not a one-time task. ERP systems are extremely complicated, and no company takes full advantage of all the capabilities in the system. Firms that implemented ERP systems to avoid the Y2K problem likely installed ERP systems that only covered the basic functionality necessary to operate the business through the Y2K transition. Many of these firms are now looking to leverage their ERP systems to improve their business processes, which means that their implementation projects, while smaller than the initial ERP implementation, still require effective management.

Implementation Choices for the Small to Midsized Market

Traditionally, when small to midsized companies looked to implement an ERP system, they would buy one that was offered regionally and for their specific industry; many European companies adopted this strategy because the European ERP vendors were familiar with regional differences in business. However, now that the marketplace and manufacturing are global scenarios, this method of choosing ERP systems is no longer valid. Further, these smaller, regional ERP vendors often offered only one part of the package. If a company wanted more, it had to piece together different ERP systems.

Now the small to midsized market is being served by the larger ERP vendors, in an environment that is experiencing tremendous consolidation. Companies are concerned that if they choose an ERP vendor, that vendor could be bought by another vendor that would provide minimal update support for the original ERP system. Or in some cases, the company might have to migrate to a different ERP system. For example, PeopleSoft, once an independent vendor, is now one of several ERP products owned by Oracle. Oracle also has purchased J.D. Edwards, Retek, and Siebel. In 2005, when Oracle took over PeopleSoft, it laid off 5,000 employees but spared staffers directly involved with customers and with code and technology originating from PeopleSoft and J.D. Edwards.

The two largest players in the ERP market, SAP and Oracle, are now focusing on small to midsized customers (companies with fewer than 1,000 employees). These two vendors enjoy a direct advantage over their smaller competitors because of their huge research and development efforts. The third-largest ERP vendor is a company called Sage Software, which has acquired a number of companies, including the accounting software manufacturer Peachtree. The fourth player in this market is Infor, which also has acquired a large number of companies, including MAPICS (with 4,500 customers, including Volvo Construction Equipment) and Mercia Software (which services General Motors and Coca-Cola). Microsoft, which acquired Great Plains, Solomon Software, and Navision, is growing at 20 percent per year with revenues exceeding $800 million, and is estimated to be about fifth in the ERP market.

One very important criterion for a small or midsized company to identify when selecting an ERP system is the depth of the system. The ERP software must be capable of handling the industry that the company is in, with all of its fine details. For example, an ERP system that specializes in accounting will not be very useful if you are running a law firm. SAP addresses this need for specialization by partnering with firms that can provide that industry depth—or, as it is known, serve a "vertical." SAP's product, All-in-One, is created with industry partners to provide a single solution for a specific industry. By partnering with other firms who have expertise in the specific areas, SAP is able to offer many more tailored packages.

Questions:

1. What specific needs do you think a small to midsized company has in selecting an ERP vendor? How do these needs differ from those of a large company?
2. If your ERP vendor is bought by another, what challenges might you face in terms of your ERP system?

ERP System Costs and Benefits

As you learned in Chapter 2, ERP implementation is expensive (usually ranging between $10 million and $500 million, depending on company size). Among its costs are:

- Software licensing fees: ERP software is quite expensive, and most ERP vendors charge annual license fees based on the number of users.
- Consulting fees: ERP implementations require the use of consultants with detailed knowledge of how to configure the software to support the company's business processes. Good consultants have extensive experience in the way ERP systems function in practice, and they can help companies make decisions that avoid excessive data input, while capturing the needed information to make managerial decisions.
- Project team member time: ERP projects require key people in the company to guide the implementation. These team members have detailed knowledge of the company's business, and they work with the consultants to make sure that the configuration of the ERP software will support the company's needs. This means that these workers are frequently removed from their daily responsibilities to work on the implementation project.
- Employee training: Project team members need training in the ERP software so that they can work successfully with the consultants in the implementation. The team members frequently work with training consultants to develop and deliver company-specific training programs for all employees.
- Productivity losses: No matter how smooth the ERP implementation, companies normally lose productivity during the first weeks and months after switching to the new ERP system.

Companies must identify a significant financial benefit that will be generated by the ERP system, to justify the money they will spend on it. The only way companies can save money with ERP systems is by using them to support more efficient and effective business processes. This means that implementation projects should not re-create the company's current processes and information system in a new ERP package—which is a very real possibility. SAP provides all source code with its ERP package, which means that the user can see how its ERP system is designed and can alter the package through its internal programming language, called Advanced Business Application Programming (ABAP), which you first learned about in Chapter 2. With access to the SAP ERP source code, it is possible for a company to spend a significant sum of money on software code development to avoid changing its business process to the best practices designed into the ERP software. Recall that many companies would prefer to avoid changing their processes and continue doing business as they always have – rather than adopt the best practices built into the ERP system.

Finally, companies must manage the transfer of data from their old computer systems to the new ERP system. In addition to managing master data such as materials data, customer data, vendor data, and so on, a company must also transfer transaction data, which includes sales orders and purchase orders, many of which are in various stages of processing—a challenging task.

IMPLEMENTATION AND CHANGE MANAGEMENT

How does a company make sure that its ERP investment pays off in increased profitability? The key challenge is not in managing technology, but in managing people. An ERP system changes how people work, and for the system to be effective, the change may have to be dramatic, going beyond the look of the software and into the way employees perform their tasks. Furthermore, business processes that are more effective require fewer people. Some employees will no longer be needed. It is no small thing to ask people to participate in a process that may not only change their day-to-day activities, but could also eliminate their current jobs.

Managing the human behavior aspects of organizational change is called **organizational change management (OCM)**. Do not underestimate the importance of this part of the implementation process. One of the keys to managing OCM is to realize that people do not mind change, they mind *being* changed. If the ERP implementation is a project that is being forced on the employees, they will resist it. If employees view it as a chance to make the company more efficient and effective by improving business processes, and if these process improvements will make the company more profitable and therefore provide more job security, then there is a greater likelihood that employees will support the implementation efforts. As the previous section mentioned, the best way to improve a business process is to have the people most familiar with the process participate, using their experience and creativity to develop process improvement ideas. When employees have contributed to a process change, they have a sense of ownership and will likely support the change.

Implementation Tools

Many tools are available to help manage implementation projects. Process mapping, described previously, is perhaps the most critical. For an ERP implementation to go smoothly and provide value, it is critical that a company understand both its *current* processes and the state of the process *after* implementation.

SAP provides Solution Manager, a tool that helps companies manage the implementation of SAP ERP. In Solution Manager, the ERP implementation project is presented in a five-phase Implementation Roadmap. The five phases are:

1. Project Preparation (15 to 20 days)
2. Business Blueprint (25 to 40 days)
3. Realization (55 to 80 days)
4. Final Preparation (35 to 55 days)
5. Go Live and Support (20 to 24 days)

Figure 7-17 shows an example of this roadmap. The left side of the Solution Manager screen shows a hierarchical menu structure that organizes each step in the process, and on the right side of the screen are the detailed items and explanations for one step.

Solution Manager has tools to support each phase in the roadmap, including documents, reports, white papers, and planning tools. The first phase of the Implementation Roadmap is Project Preparation. Some of the tasks in Project Preparation include organizing the technical team, defining the system landscape (including servers and network), selecting the hardware and database vendors, and, most importantly, defining the project's

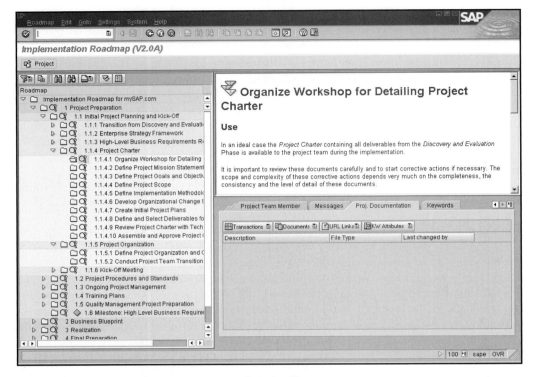

FIGURE 7-17   Implementation Roadmap in Solution Manager

scope—what the project is to accomplish. A common problem in ERP implementations is **scope creep**, which is the unplanned expansion of the project's goals and objectives. Scope creep causes the project to go over time and over budget, and increases the risk of an unsuccessful implementation. Defining the project's scope ahead of time helps prevent this problem.

The second phase, the Business Blueprint, produces detailed documentation of the business process requirements of the company. The Business Blueprint provides a detailed description of how the company intends to run its business with the SAP ERP system. Process mapping is critical in the Business Blueprint phase. The Business Blueprint guides consultants and project team members in configuring the SAP ERP system (which occurs in the third phase). During the Business Blueprint phase, technical team members determine the method of data transfer from the firm's existing computer systems (called legacy systems), which will either be replaced by the ERP system or will continue to function with the ERP system through an interface.

In the Realization phase (the third phase), the project team members work with consultants to configure the ERP software in the development system. The team also develops any necessary ABAP code or other tools (such as third-party software packages) and creates the required connections to the legacy systems.

The fourth phase, Final Preparation, is critical to the success of the implementation project. Tasks in this phase include:

- Testing the system throughput for critical business processes (determining whether it can handle the volume of transactions)
- Setting up the help desk where end-users can get support
- Setting up operation of the Production (PROD) system and transferring data from legacy systems
- Conducting end-user training
- Setting the Go Live date

When scope creep occurs in a project, it is commonly not discovered until well into the Realization phase, when the team begins to miss deadlines, and the costs begin to exceed the budget. By the time the scope creep is discovered and its impact is understood, there is little management can do to correct the problem, as most of the time and budget have been spent. Management can choose to shorten or omit the Final Preparation phase, which means that testing of the system and training of employees are reduced or eliminated. Unfortunately, with reduced testing, errors in configuring the system are not discovered until it is put into use. Likewise, with reduced training, employees do not know how to use the system properly, which can create a complicated chain of problems, due to the integrated nature of the system. Any cost savings gained by shortening the Final Preparation phase are overshadowed by productivity losses and consulting fees in the Go Live and Support phase.

In the fifth and final phase, Go Live and Support, the company begins using the new ERP system. Wise managers try to schedule the Go Live date for a period when the company is least busy. Setting up a properly staffed help desk is critical for the success of the Go Live phase, because users have the most questions during the first few weeks of operating with the new system. The SAP ERP project team members and consultants should be scheduled to work the help desk during the first few weeks of the Go Live period. Although significant testing of the system and settings should have been done throughout the project, it is not possible to test all the settings and thoroughly evaluate the throughput of the system. Therefore, monitoring of the system is critical so that changes can be made quickly if the performance of the system is not satisfactory. Finally, it is important to set a date at which the project will be complete. Any enhancements or extensions to the system should be managed as separate projects, not as extensions of the original implementation project.

Training Challenges

In 2007, Kimberly-Clark, the $17 billion maker of disposable diapers, tissues, and toilet paper, completed a $100 million implementation of SAP in 32 manufacturing mills in North America. Installing the software for the business processes of requisition to check, accounting to reporting, plant maintenance, and materials management was enough of a challenge. But the real test came in the training of Kimberly-Clark's employees. It cost the company $17 million to train all of the 16,000 employees in the affected facilities. Most of the employees being trained were mill workers, some of whom had never used a computer. So basic skills like how to use a mouse were required in the training sessions.

Most of the mill jobs are not high-tech positions, so it may seem strange that installation of the SAP system has changed the way mill workers do their jobs. For example, if a part on a machine breaks down, the mill worker used to call the parts department to report the broken piece. The worker in the parts department would take the information down on paper. Now, the mill worker reports a broken part on the SAP system. Management can now track which parts are breaking down and how quickly they are being fixed, and can compare those data to those of other mills.

Kimberly-Clark is also tracking usage of the SAP system and spotting troublesome areas, using specialized technology called "IT end user experience monitoring software." Kimberly-Clark uses Knoa Experience and Performance Manager for SAP, which helps the company understand how users interact with the system. The Knoa software will send an alert if the system is slow in one area, which can indicate that users are having trouble with this area. By issuing reference cards to help employees with one trouble spot, Kimberly-Clark saved $395,000 on additional training. Through this software, Kimberly-Clark can ascertain whether the problem is user-based or system-based. Forrester Research predicted that the market for this type of software would grow 20 percent by the end of 2007 to $138 million.

Kimberly-Clark is planning to implement more modules in the SAP system in 2008.

Question:

1. Research IT end user experience monitoring software and report on the details of this software.

System Landscape Concept

SAP recommends a system landscape for implementation, like the one shown in Figure 7-18. In this system landscape, there are three completely separate SAP systems, designated as **Development (DEV)**, **Quality Assurance (QAS)**, and **Production (PROD)**. The Development (DEV) system is used to develop configuration settings for the system, as well as special enhancements using ABAP code. These changes are automatically recorded in the **transport directory**, which is a special data file location on the DEV server. These changes are imported into the QAS system, where they are tested to make sure that they function properly. If any corrections are needed, they are made in the DEV system and transported to the QAS system. Once the configuration settings and ABAP programs pass testing in the QAS system, all settings, programs, and changes are transported to the PROD system, the system that the company uses to run its business processes.

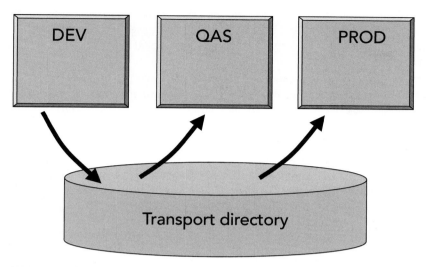

FIGURE 7-18   System landscape for SAP ERP implementation

The use of separate systems is important during the initial implementation of an SAP ERP system, and it is even more important after the Go Live phase. All software packages have occasional updates, and having systems available to test these updates before applying them to a production system can prevent problems. If a company wishes to use features of the SAP ERP system that were not included in the initial project implementation, then the company should have a process like the one SAP provides to manage changes to the production system in a controlled fashion.

Chapter Summary

- The concept of business processes has been an important theme throughout this book. ERP systems are designed to provide the information, analysis tools, and communication abilities to support efficient and effective business processes. This chapter introduced process modeling as a fundamental tool in understanding and analyzing business processes.

- Process mapping is one process-modeling tool that uses graphical symbols to document business processes. Other methodologies include hierarchical modeling, deployment flowcharting, event process chain diagramming, value analysis, and business process improvement. SAP's Solution Manager, a set of tools and information that can be used to guide an implementation project, is included in SAP ERP to help manage the implementation of the ERP software.

- SAP's system landscape was introduced to show how changes to the ERP system during implementation (and beyond) are managed.

- Most challenges to ERP implementation involve managing personnel and their reactions to the change, rather than managing technical issues.

Key Terms

Business process innovation (BPI)

Business process reengineering

Continuous improvement

Deployment flowcharting (swimlane flowcharting)

Development (DEV) system

Dynamic process modeling

Event process chain (EPC)

Flowchart

Gap analysis

Hierarchical modeling

Organizational change management (OCM)

Process boundary

Process mapping

Process model

Production (PROD) system

Quality Assurance (QAS) system

Quality circles

Scope creep

Transport directory

Value added

Value analysis

Workflow

Workflow tasks

Exercises

1. Develop a process map for the process a professor must follow to prepare a multimedia presentation of a concept for students. Developing a multimedia presentation can require various steps:

   a. First, if the concept is not abstract, then it may be possible to record the subject.

   b. If you have access to the subject of your multimedia presentation, then you should record audio, video, or graphic images, as appropriate.

   c. If your concept is abstract or you do not have access to it, then you should search to see if someone has already created media that illustrate the concept sufficiently for your use.

d.   If an alternate source does not exist, you must create drawings, animations, or simulations of the concept.

e.   If you find existing material, you have to obtain copyright clearance, unless it is in the public domain or your use is covered by the Fair Use Guidelines (check with your instructor on this).

f.   If the media are not in digital form, you must digitize them before distribution to students.

2.   Develop a process map for the process of handling food shipments at a FoodMore Grocery Store:

*   If a truck that contains perishable items (such as dairy, meat, or produce) arrives at a FoodMore store, the temperature of the truck has to be checked to ensure that the items were transported properly. Each product category has a product data sheet that defines the acceptable temperature range for the truck. If the truck is not in the proper temperature range for the product category, then all items of that category are rejected. More than one product category may be on a truck if the temperature ranges for the products are compatible.

*   As cases are removed from the truck, they are checked for damage. Damaged cases are rejected. The expiration dates are checked against the date information on the product data sheets to make sure each case will have an acceptable minimum shelf life. If a case has an unacceptable expiration date, then the case is rejected.

*   Differing processing steps are followed for all accepted cases, depending on whether they are perishable. Once a case of perishable food is accepted, it is moved immediately to the cooler to await further processing. Some products, like beef, pork, whole chicken, and cheese, are bought in bulk and must be cut and packaged before being put on the shelves. The processing procedures are defined in the product data sheets for each food. Bulk foods are processed on an as-needed basis, and a stock clerk takes the processed packages to the appropriate location in the store immediately after they are processed and packaged. The stock clerk also provides one last check of package integrity before placing the items on the shelves and display cases. If the package is damaged, the item is discarded.

*   If an item is not perishable, it is placed in the storeroom and then moved onto a store shelf when the number of items on the shelves gets low. The stock clerks also check the package integrity of the nonperishable items before placing them on the shelves, looking for damage such as dents in cans. Depending on the level of damage, dented cans may be marked down in price and placed on the shelves in a special reduced-price area. The clerks might discard the items if the damage is too great.

*   Once items are on the store shelves, they must be monitored periodically to make sure that the expiration dates have not passed. If an item is getting close to its expiration date, the price may be marked down for quick sale. If the expiration date has passed, the item must be discarded.

3. Develop a process map for the following process, which describes the sales process for Daisy Scout cookies. The cookies generate income for individual scout troops and for the regional scout council. Use a deployment (swimlane) format. Can you suggest any changes to improve the process?

    *   The Daisy Scout troop leader for each troop receives the order forms from the local Daisy Scout council office. The leader distributes the order forms to the girls at the beginning of the cookie campaign. The girls go out and book cookie orders during the first order period. If a scout completes the entire form, she can turn it in to the troop leader and get another form. At the end of the first order period, the scout leader collects all the cookie order forms, makes copies, and gives the forms to the "cookie mom" (a volunteer who manages the cookie orders for the scouts). The cookie mom tabulates all of the orders and then delivers a consolidated order form to the Daisy Scout council cookie coordinator. The troop order must be for full cases (12 boxes per case). The troop's orders might not total full cases, meaning that it will receive more cookies than the scouts sold.

    *   When the cookie manufacturer delivers the cookies to the Daisy Scout council office, the council contacts the cookie mom, and she picks up the cookie order from the council office, checking the delivered quantity against the order. The cookie mom takes the order home and then breaks the cases down into groups for the individual orders. She delivers these orders to the girls at their next meeting. The girls deliver the cookies to the customers, collecting $3 for each box. The girls return the money to the Daisy Scout troop leader, who passes it on to the cookie mom. The cookie mom deposits a portion of the money in the troop's account and deposits the council's portion in the council bank account. This process must be documented at the bank and all receipts turned in to the council before more cookies can be ordered. The portion of the sales revenue that the troop keeps depends on the sales level. Normally they collect $0.50 per box, but if the troop sells an average of 100 boxes per Daisy Scout, they collect $0.55 per box. The council also gets a percentage of the proceeds to support their operations. The troop can place a second order with the council coordinator if they deposit by the required date at least 50 percent of the money they collected from their first sales effort.

    *   The second order also has to be for case quantities, and the excess boxes from the first order must be considered. To avoid buying an entire case of cookies to get one or two boxes needed for the orders, the cookie mom can deal with cookie moms from other troops to buy excess boxes of cookies (although this practice is discouraged by the council office). When buying a box of cookies from another cookie mom, the moms have to agree on which troop gets to keep the profits. This cookie dealing can be reduced if the troop decides to have a direct sales effort. For example, they might set up a table at the local grocery store and sell boxes of cookies for cash. If the troop wants to do this, they have to get approval from the council office. The council office assigns the selling locations, so the troops should do this early in the process, to get their choice of location. If her troop is holding a direct sales event, then the cookie mom can dispose of any excess boxes from her own troop's orders, although she may choose to help out other cookie moms by selling boxes to cover some of their orders (again, the sales price has to be determined).

    *   Once all the money from the second round of orders and the direct sales is collected, the cookie mom needs to deposit the money in the troop and council accounts. She must

submit all paperwork to the council cookie coordinator by a given date. It is possible that the troop sold less than 100 boxes per girl in their first effort but achieved that goal with the second order and direct selling. In that case, the troop gets to keep $0.55 for all boxes sold, and the deposit needs to be calculated accordingly. Any leftover boxes can be used for snacks at the troop meetings, and are paid for with troop funds.

4. For the accounts payable process described below:

   a. Develop a process map of the current process.

   b. Analyze the value added by each step in the process. Determine whether the step adds real value, business value, or no value.

   c. Develop a process map for an improved process.

- The accounts payable process at Frizz Master Hair Product Manufacturer begins when specific departments ask to purchase certain products or equipment. The department manager then approves the requests so that the buyers in the Purchasing department may purchase the requested materials or products.

- Once the necessary product has been bought and the invoice has been generated, an invoice for payment is sent to the Accounts Payable (A/P) department. The A/P tasks are divided between two people. One person handles invoices for companies with names that start with the letters A through M, while the other person handles invoices for companies with names that begin with N through Z. As each A/P worker opens the invoices, he or she sorts them according to the department responsible for the payment. All invoices that involve overhead (and therefore do not get directed to a specific department), such as utility and shipping invoices, are coded by the A/P clerk and entered into the accounting system for payment. All other invoices are sent by intercompany mail to their respective departments. Invoices for amounts under $1,000 are directed to the person who requested the purchase, while invoices over $1,000 are sent directly to the manager in charge of the specific department where the request originated. Once the originator or manager approves the invoice, it is sent back to the A/P clerk, who enters it into the accounting system.

- The A/P clerk selects all approved invoices that have been entered into the accounting system and prints them as a list for the accounting manager, who marks the list to show which invoices are to be paid during the current check cycle. The accounting manager gives the annotated invoice list back to the A/P clerk, who reports to the staff accountant to retrieve a special disk that allows access to the check-printing system. The disk must be kept under lock and key, which is the responsibility of the staff accountant. The A/P clerk uses the disk to connect to the check-printing system and print checks to be immediately mailed to the vendors, along with the corresponding invoices. The A/P clerk must keep written records of the check numbers and the amount of each check that is printed and mailed. Finally, the account manager double-checks for errors and signs off on the check number and dollar amount of each check, after each check run.

5. Develop an EPC diagram for the following Human Resources process:

   a. The current recruitment process for Yellow Brook Photography takes approximately 90 days. It begins when a manager has a need for a position. The manager must complete a requisition and send it to the Human Resources (HR) department. HR reviews and assigns a number to the requisition, and returns it to the manager for approval. He approves it, obtains the appropriate signatures, and then returns it to HR.

b.   Next, HR creates a job posting and announces the position internally, first through the company's intranet, bulletin boards, or a binder of current job openings. HR collects responses for eight days. HR also solicits resumes from external sources by advertising. HR prescreens resumes and forwards data on qualified candidates to the hiring manager for review. The hiring manager notifies HR of candidates to interview. She also conducts phone screens; if the phone screen is promising, HR coordinates and schedules an on-site interview. Candidates interview with the hiring manager and with HR. HR records the interviews in an applicant flow log.

c.   Once a candidate is selected for hire, HR and the hiring manager prepare an offer, and the background check is initiated. The hiring manager then must approve the offer and obtain the required signatures on an internal associate data/change form. Subsequently, she must extend the offer verbally to the candidate, while HR sends the written offer, including a start date for work. Once the applicant accepts the offer, a drug screening is scheduled with the candidate, who must also sign the offer letter and return it to HR. HR notifies the hiring manager of the candidate's acceptance. Finally, if the drug test comes back negative, the new employee completes "new-hire" orientation on the date hired.

For Further Study and Research

Cordes, Ronald M. "Flowcharting: An Essential Tool." *Quality Digest*, January 1998.

Cowley, Stacy. "Update: Oracle laying off 5,000." *Computerworld*, January 14, 2005.

Hammer, Michael, and James Champy. *Reengineering the Corporation: A Manifesto for Business Revolution.* HarperCollins. 1993.

Harrington, H. James. *Business Process Improvement: The Breakthrough Strategy for Total Quality, Productivity, and Competitiveness.* New York: McGraw-Hill, April 1991.

MacIver, Kenny. "ERP roulette." *Information Age*, July 11, 2007. http://www.information-age.com/article/2006/february/erp_roulette.

CHAPTER 8

ERP AND ELECTRONIC COMMERCE

LEARNING OBJECTIVES

After completing this chapter, you will be able to:

- Describe business-to-business e-commerce.
- Explain the importance of ERP to the success of a company engaged in e-commerce.
- Describe the function of an application service provider (ASP).
- Describe the delivery of ERP services through an ASP.
- Describe Web services and SAP's NetWeaver.
- Describe the unique components of NetWeaver.
- Explain why accessing an ERP system through a Web browser is efficient.
- Define XML and its significance to ERP.
- Define RFID and its future role in logistics and sales.

INTRODUCTION

This chapter examines connectivity topics relevant to Enterprise Resource Planning (ERP) systems. As you have read, an ERP system lets a company accomplish things that cannot be done well, if at all, without such a system. In this chapter, you'll learn how effectively competing in high-volume e-commerce may be impossible without the infrastructure provided by ERP. You will further learn how companies can integrate ERP systems with the Internet and "rent" ERP software from special-purpose software companies. This chapter also explains the basic components of Web services, with a focus on NetWeaver, SAP's Web services platform. You will find out why XML is becoming the new markup language of the Internet and how

ERP companies are using it. Finally, you will read about radio frequency identification (RFID) devices and their significance to managing the movement of goods in the supply chain.

ELECTRONIC COMMERCE BACKGROUND

Today, most companies conduct at least part of their business operations through electronic commerce, more commonly known as e-commerce. As you learned in Chapter 2, e-commerce is the conduct of business over the Internet. When people think of e-commerce, they often think of retail e-commerce, typified by companies such as Amazon.com. Most of the business growth on the Internet, however, has been in the area of **business-to-business (B2B)** e-commerce, rather than in retail **business-to-consumer (B2C)** e-commerce.

Business-to-Business E-Commerce

Business-to-business e-commerce is defined as buying and selling between two companies over the Internet. The companies might be manufacturers, suppliers, wholesalers, or retailers.

Business-to-business e-commerce is transforming the way companies work with each other. For example, industry-specific online auctions allow companies to buy raw materials at the lowest possible cost. Say a food company's purchasing manager wants to buy honey. She can go to an auction site—either for the food industry or a general "wholesale/retail only" auction site—and get competitive bids simply by posting the request rather than requesting competitive bids from each supplier independently by e-mail or fax.

Electronic Data Interchange (EDI)

B2B e-commerce is not new. Prior to the development of the Internet, companies electronically transferred purchase orders through a system known as electronic data interchange (EDI). Recall from Chapter 2 that EDI is an electronic computer-to-computer transfer of standard business documents. Companies have been doing EDI since the 1960s and originally used telephone lines to transfer data.

Companies can set up their own private EDI networks to communicate directly with their suppliers' systems. EDI networks are very expensive, so many companies subscribe to a **value-added network (VAN)**, an intermediary Internet-based network run by an outside EDI service provider. With EDI, when a company needs to order raw materials, it can place the order electronically. EDI should not be confused with a simple e-mail message between two parties that reads, "Dear Joe, please send us 500 gallons of honey." In EDI, the order is sent on a standardized business transaction form, following a specific computer protocol.

The benefits of EDI are enormous:

- Costs of paper, printing, and postage have almost disappeared from ordering systems.
- Errors have been minimized because orders are not manually entered into the supplier's information system.

- Ordering is fast and efficient. As a result, large companies can force their smaller suppliers to use EDI, through a VAN if necessary. The supplier pays the VAN's fees, based on the amount of information sent and received.
- Suppliers and buyers are "locked" into business relationships: Once a company sets up an EDI system with its supplier, changing suppliers becomes a major inconvenience. Most buyers don't change suppliers, hence locking in the relationship. This is an advantage for suppliers and buyers, as long as both remain satisfied.

Internet-Based Procurement

Even though EDI has been useful, companies are moving from EDI to **Internet-based procurement**, which is the use of Internet technologies for procurement activities (recall that procurement is the buying of raw materials for manufacture, or purchase of finished goods for resale). Internet-based procurement provides the following benefits:

- It is less expensive to use the Internet than private EDI networks.
- Purchasing costs are further reduced as suppliers compete for orders on the buyer's Web site.

Locking in suppliers often does not occur in Internet-based procurement. Buyers tend to ignore suppliers who cannot compete on price, instead focusing on relationships with viable suppliers. Smaller companies now can get a share of the market by connecting electronically with larger customers. Vetco International, Inc., a small supplier of equipment in the petroleum industry, can count giants such as Exxon Mobil Corp. and British Petroleum PLC as customers because the petroleum companies link their product catalogs electronically. Vetco spent about $150,000 in 2004 to implement SAP's Business One suite, which allows the company to connect easily with its customers' SAP ERP systems.

Buying and selling goods on the Internet has evolved into the concept of electronic marketplaces. To understand this concept, think about how farmers' markets work. Farmers, growers, and gardeners bring their produce to a central location in town, where customers can compare sellers' goods on quality and cost. It is efficient for both buyers and sellers. Being grouped together, sellers gain access to more customers than they would on their own. Customers can shop for a variety of items in one place, making it easier to find the best quality at the lowest cost.

An **electronic marketplace** is a gathering place for buyers and sellers on the Internet. It can be run with or without a central operator. Marketplaces without central operators are simply lists of Web sites directing buyers to certain products. Marketplaces with central operators facilitate and expedite the buying and selling of goods. Businesses using this type of marketplace may have to pay fees to join, and they usually pay fees to use the marketplace. **Exchanges** are one type of B2B electronic marketplace. Exchanges typically focus on a single industry.

Chemical buyers, sellers, and traders can find their products at the ChemConnect marketplace, www.chemconnect.com. The Web site claims the following advantages:

- Buyers can find the best prices without traditional negotiations.
- Faster contracts are completed between buyers and sellers.

- Buyers and sellers can gain access to new worldwide markets and new trading partners.
- Instant market information is available to all parties.

Members of the ChemConnect marketplace use the marketplace exchange to react to changes in the often-volatile chemical market faster than they could using more traditional methods of buying and selling. For example, Vanguard Petroleum Corporation places bids and offers simultaneously on ChemConnect to multiple customers. About 15 percent of Vanguard's spot purchases and sales of natural gas liquids are conducted on the ChemConnect site. Vanguard can now publicize its offers to 150 potential partners in only a few minutes; prior to its involvement with the exchange, it took an entire day to get the information out to that many companies.

Another type of industry marketplace is the private exchange, where membership is restricted to select participants. Private exchanges can be attractive to businesses because they are limited to trading partners with which the responsible party has an established relationship. For example, IBM runs a private exchange for its customers and suppliers; IBM's competitors are prohibited from participating in the exchange. Volkswagen has slashed procurement costs by up to half through its use of a private exchange.

Internet Auctions and Reverse Auctions

As previously mentioned, B2B e-commerce allows companies to do online bidding through auctions and reverse auctions. **Reverse auctions** feature one buyer and many sellers. A company can use a standard auction to put its products or even its obsolete equipment up for bid, and at the same time use reverse auctions to request bids from suppliers for goods or services.

For example, Fitter Snacker needs raw materials such as oats and wheat germ, which for FS are essentially commodities. **Commodities** are items that are widely available at a standard level of quality; the only thing that varies between one commodity and another is price. The company can go to a "bidding" Web site—one that seeks bids from sellers rather than buyers—and set up a reverse auction program to run overnight. The program uses the Internet to solicit bids to supply those raw materials. In the morning, FS's purchasing agent can choose the lowest price bid for the oats and wheat germ.

Internet-based auctions are changing the way in which commodities are purchased. A few years ago, commodities would have been purchased through a supplier, or intermediary, who negotiated prices of raw materials with sellers. Now, the Internet and its bidding programs have threatened the intermediary's role and made the buying process more efficient. In a sense, the Internet has replaced the intermediary. Pricing is open and dynamic, meaning that competitors can see each others' bids (although they probably can't see who is making the bids). Epsilon Products Company, a producer of polypropylene, has used a reverse auction on the ChemConnect marketplace to reduce its cost by 5 percent.

Dynamic pricing is not only forcing out intermediaries, it is also putting pressure on sellers to be flexible. This means that a seller's accounting and logistics operations must be in excellent shape (requiring an ERP system) before it tries to sell in the auction market, as competition is based on price.

Electronic Commerce Security

E-commerce brings with it a major concern: security. No company (or individual, or government agency, for that matter) is immune to security breaches. In the past, large firms have been shut down due to various types of systems attacks, including denial of service attacks. In **denial of service (DoS)** attacks, attackers block access to a Web-based service through a variety of means, including bombarding a site with so many messages that the site can't handle the volume.

On February 6, 2007, a denial of service attack was launched against DNS servers, the computers that handle the domain name service of the Internet, in charge of Web page names. Six of the root servers were flooded with bogus queries at the rate of 1 GB per second—roughly equivalent to 13,000 e-mails per second. E-commerce companies use virus-scanning software, encryption, intrusion detection, and other measures to protect their networks, their Web sites, and the privacy of customer data.

E-COMMERCE AND ERP

You might ask yourself, "What does e-commerce have to do with ERP?" The answer is that each technology complements the other, and each is necessary for success. An offline company cannot compete with companies offering similar goods and services if the competitors also provide the added convenience of doing business over the Internet. Without ERP, a company cannot fill orders—either Web orders or traditional orders—expeditiously. Here's why.

When a company receives an order through its Web site, the company should not merely file or print orders for later handling. The orders should be efficiently fed into the company's marketing, manufacturing, shipping, and accounting systems—a series of steps sometimes called **back-office processing**.

An efficient back-office operation is crucial for any company's success. E-commerce often exacerbates problems and reveals weaknesses in current back-office systems. Amazon.com invests its profits back into warehouses and other support that the company needs to keep its back office in order, supporting its continued success. Some Web-only businesses are worried by the entry of massive retailers like Wal-Mart into their Web marketplaces. Large companies already have well-established, integrated back-office and distribution systems, which are typically more expensive to develop than e-commerce systems, giving the large retailers speed-to-market and financial advantages.

Some companies with unintegrated information systems have built Web sites before creating an integrated back-office system. As a result, those companies often can't fill orders in a timely fashion. Many e-commerce businesses suffered from a lack of integration during the 1999 holiday season. The online toy retailer eToys.com announced less than a week before Christmas that it would *not* be able to fill all its Web orders. Surprisingly enough, all the toys were in the warehouse, but the company couldn't organize basic warehouse functions such as picking, packing, and shipping to get toys to consumers on time. Integrating the Internet front-office operation and the ERP back-office operation is fundamental in today's business environment.

Fitter Snacker and E-Commerce

Fitter Snacker has traditionally sold its NRG bars through the sales force in either the Direct Division or the Distributor Division. Now FS's customers want to order directly from a Web site. Currently, Fitter Snacker has neither a Web-based ordering system nor an ERP system. In the current annual-planning cycle, FS executives are looking at two IS investment options: (1) implement a Web-based ordering system, or (2) implement an ERP package. Company executives would like to make both investments, but they are reluctant to make two large investments in one year because of the cost and disruptive nature of the implementation efforts. Implementing an ERP system would improve its business processes, but the company is experiencing significant customer pressure for Web-site ordering.

Suppose FS executives succumb to customer pressure and implement a secure Web site, allowing customers to place orders online. FS does not know exactly what percentage of customers would use the system (although if FS already had an ERP system, that data would be readily available). To promote sales and be competitive with other online merchants, the company promises delivery from Web site orders in five working days, which seems reasonable, based on FS's past performance.

After the Web site's implementation, orders come in more frequently, which is good—it means that business is thriving. However, orders now have shorter lead times because, without an ERP system, Internet orders are saved to a text file, printed, and then manually transferred into the order entry system described in Chapter 3. The order is then handled using the warehouse procedures discussed in Chapters 3 and 4. The increase in orders is putting a strain on the company's back-office operations.

A customer who orders NRG bars from the Web site *thinks* he will be receiving the bars within five business days, as stated on the company's Web site. After 24 hours have elapsed, however, the customer receives an e-mail advising him that the NRG bars are out of stock and will not be available for at least one week, maybe two. The company's unintegrated information system, barely adequate for personal and phone-based ordering, is simply too cumbersome to handle the Internet orders. With e-commerce, customers have increased service expectations, and FS's unintegrated system cannot meet those expectations.

Now consider an alternate scenario. If Fitter Snacker implemented an ERP system first, and later connected it to a Web-based ordering system, Internet orders could flow directly into the ERP system. As customers placed orders through the Web system, the ERP system could provide accurate delivery date quotes. With this information, customers could make an informed choice and would be more satisfied. For FS, accurate and timely order data would support more effective production planning. And of course this discussion doesn't begin to address all the other ERP-related benefits that FS would enjoy.

Recent studies on back-office systems concluded that an attractive Web site does not provide enough benefit on its own for an e-commerce business to stay afloat. The conventional back-office systems must be in place and operating correctly for the business to flourish. As with any kind of business, effective infrastructure is key for e-commerce success.

Exercise 8.1

1. Assume you are opening a small business to sell a product that interests you. Some ideas might be clothing for your favorite sport or some new toys for your dog. Also assume that you only will have an Internet presence with this business—no physical store. What do you think is involved with setting up an e-commerce business? Use the Internet or interviews with people you know who have tried this. List the steps involved. Report your findings to the class.

2. Do you think customers' expectations are different when ordering on the Web versus ordering in a traditional store? Use some of your own experiences and those of your classmates to answer this question.

3. Assume Fitter Snacker implemented both an ERP system and a Web ordering system. Develop a process map of the steps needed to fill a Web order. See Chapter 7 for details on process mapping. Be sure to include a description of how the ERP system interfaces with the Web page.

4. Develop a plan for an e-commerce business that sells concert gear such as t-shirts, hats, and pins. How will it establish a competitive advantage? What role will back-office systems play?

USING ERP THROUGH AN APPLICATION SERVICE PROVIDER

Many companies today outsource some of their operations to an outside service provider, sometimes called a third party. In the next section, we will look at how outsourcing can simplify the management of ERP systems.

Application Service Providers

An **application service provider (ASP)** is a company that provides management of applications for a company over a network. Usually that network is the Internet. Companies can outsource their e-mail software, accounting programs, or other programs to an ASP.

The ASP, not the company using the ASP's services, owns the hardware and the rights to the software; it also employs the workers who run the outsourced applications. The users of the system, of course, are the company's employees. Suppose an ASP were to run an ERP package such as SAP ERP for Fitter Snacker. The ASP would own the rights to use the SAP software, as well as the server on which the SAP software runs. The ASP could also provide consulting services for configuring the ERP system, so that Fitter Snacker would not have to find its own ERP specialists. The ERP program would be delivered to FS workers over a network, probably the Internet. The ASP would charge a monthly or yearly fee, or it could charge a per-use fee for the system, based on the number of users at FS. Thus, an ASP can provide ERP software with a much lower start-up cost, making it possible for smaller companies to use ERP systems when they cannot afford the costs of installing and maintaining their own system. Figure 8-1 outlines the details of running an ERP system in-house, versus using an ASP.

Fitter Snacker in-house ERP

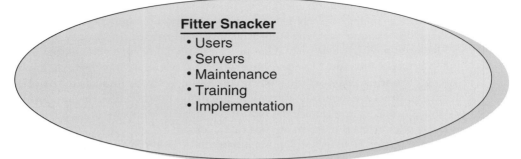

Fitter Snacker ASP ERP

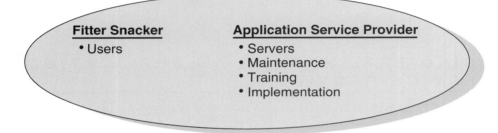

FIGURE 8-1   ERP responsibilities in-house versus with an ASP

Advantages of Using an ASP

Many companies find it advantageous to use an ASP for ERP and other information system applications. Some of the benefits include:

*   Affordability: Companies that cannot afford their own ERP system can now "lease" one on a monthly basis, avoiding the high cost of obtaining the hardware and software and hiring and training support personnel. ASP services can be received through the Internet, using either a Web browser or the ERP system's graphical user interface (GUI) software.
*   Shorter implementation time: The time needed for ERP implementation is shorter for those who implement ERP through an ASP. An ASP should have experience with implementing and maintaining ERP software. (If it does not, the company should find another ASP.) ASPs have servers, telecommunications, and highly trained personnel already in place.
*   Expertise: ASPs are experts in delivering IS applications. They do all the maintenance, including execution of backups, training, and customizing of the system. Customers of ASPs do not need to hire additional IT personnel. ASPs can also run information systems more efficiently because they do it on a large scale. ASPs can spread fixed costs over many users, thus achieving economies of scale that might translate into a lower total cost of ownership. The

availability of IT talent is tight in the current market; an ASP might have a better chance of attracting and retaining a talented workforce than a small manufacturing company like Fitter Snacker.

Disadvantages of Using an ASP

There are some potential problems with using an ASP:

- Security: Companies using ASPs are turning their information systems over to a third party. They must be confident that the ASP has a high level of security. How hardware will be shared is also a security concern. The ASP will have multiple users on a single piece of hardware. Each customer's data must be shielded from other customers. While these are valid concerns, ASPs often have better security than a small company can have on its own.
- Bandwidth/response time: The telecommunications channel from the ASP to its customers must be fast enough to handle multiple users. An ASP's servers must be sufficient in terms of processing capabilities.
- Flexibility: An ASP should be flexible in working with its users and satisfying their requests for processing modifications.
- No frills: An ASP can usually provide basic ERP systems well, but asking for unusual configurations may cause problems, and an ASP might not allow for third-party add-ons. Further, the ASP might not want to do custom software development using the SAP ERP programming language, ABAP.
- Technical, not business focus: An ASP knows the technical aspects of the software, but it will need the customer to define the business processes and make the configuration decisions.

ANOTHER LOOK

Using ERP Through an Application Service Provider

SAP software is run at over 400 universities as part of SAP's University Alliance. The University Alliance program grants access to SAP's ERP software for classroom use. At the time the University of Delaware began its University Alliance program, in the 1990s, each university was required to purchase its own server (which at the time could cost upwards of $50,000) and had to train its own system administrators. With over 100 universities in the Alliance, managing all of the systems became a significant problem for SAP. On the academic side, operating the system was a chore for the university's systems administration personnel because managing the SAP system was usually an added duty on top of their normal responsibilities. To address these problems, SAP's University Alliance program developed its own ASPs, which it termed University Competency Centers (UCC). Rather than having every university administer its own system, the Alliance has four universities serve as UCCs and host the other members of the University Alliance, meaning that the UCCs act as ASPs for the other members. For SAP, the task of managing four UCCs is much easier than providing support for hundreds of individual universities. Another benefit is that the UCCs have

continued

worked together to develop unique competencies in administering systems for education—a specialized task. All universities that use a UCC to host their systems gain from this increased competency. If every university hosted its own system, it would be very difficult to share the knowledge developed at hundreds of universities.

For students, accessing an SAP system from a UCC seems no different from accessing a system at their own university, except that the UCC system is usually much faster, and the software is current and ready for use.

Questions:

1. What are the advantages for the University of Delaware of using an ASP for its SAP delivery? Are there any disadvantages?
2. What are the challenges to a university of running a complicated program such as SAP? Can you relate those challenges to small companies?

Other Considerations

As with all forms of outsourcing, companies considering an ASP should carefully scrutinize the ASP's contract to uncover hidden costs and potential problems before signing it.

ERP companies are excited about ASP capabilities because ERP vendors can provide ASP services too. Such an offering translates to extra profit and a steady income. Prior to delivering software through an ASP, ERP software companies relied on a small number of very large sales to make their yearly profit. Now income is steadier because ERP vendors can deliver and support software on a monthly basis to customers of all sizes. For example, an ERP vendor might have high revenues one quarter from securing a few, large sales, but may follow that with lower revenues the next quarter if no large software deals are finalized. Delivering software through an ASP, however, provides a steady monthly income from lease payments.

SAP is offering an ASP version of its ERP product for midsized companies. In September 2007, SAP introduced Business ByDesign, which is delivered to customers over the Web. At that time, pricing was set at $149 per month per user, and the product included applications for financials, human resources, supply chain management, and customer relationship management. Also available in a smaller package, Business ByDesign is available to companies for $54 per month per five users. SAP hopes to increase its customer base with the new product to 100,000 by 2010.

Exercise 8.2

Let's return to our example company, Fitter Snacker. Assume that under its newly appointed CIO, FS has made the decision to acquire an ERP system. Now the CIO must decide how to implement the system. After talking with the vendors of ERP software, the CIO realizes that FS has two options for implementing ERP software:

1. Buy the rights to the ERP software and any new hardware required to run it, and also hire and train system administrators.
2. Run FS's ERP system over the Web through an ASP, which would deliver ERP services for a monthly fee.

Both types of implementations have advantages and disadvantages, as you have already seen. In this exercise, you must recommend one course of action to FS's CIO. For the first part of this exercise, write a memo to the CIO enumerating the pros and cons of each method. Use the table shown in Figure 8-2 to organize your thoughts.

Advantages of purchasing software and computers for ERP	Advantages of using an ASP to run ERP

FIGURE 8-2   Arguments for purchasing software versus using an ASP

After distributing your memo to various FS board members, the CIO needs to justify his decision. He would like a spreadsheet that analyzes the financial impact of both scenarios. Your job is to compare the monthly cost of using the ASP with the cost of an in-house ERP system. Weigh the pros and cons of each method, and then make a recommendation to the chief financial officer of FS. Keep in mind that Fitter Snacker's net income is $3.4 million (you can find more details in Chapter 5). Here are the details of the decision:

Option 1: Buying Computers and Software Rights

To set up its SAP system, FS must buy:

- *Database server:* The server would cost $70,000.
- *Application server:* FS needs a server to run the ERP application, which would cost about $40,000, since there aren't any servers available for use in the system. (ERP software is platform-independent; therefore, it can be run on different types of computers. Thus, a company can often use an existing server.)
- *PCs:* Some of FS's existing PCs could be loaded with the ERP software's GUI and be used to access the system. Because more FS employees will be connected to the new system, however, FS will need 10 additional computers. Total cost for the PCs would be $15,000.
- *Rights:* Rights to the ERP software for all users for five years would cost $500,000. The CIO does not know whether further outlays will be required after the fifth year, and therefore is limiting the analysis to the years 2008 through 2012.
- *Installation:* The ERP vendor will help install the system, but FS also needs to hire consultants for the six-month implementation. At $3,000 per day, the cost is estimated at $486,000.
- *User training:* With the purchase of the rights to the ERP software, FS employees receive training at a local training center. This training is for key personnel involved in the ERP implementation project. FS wants further training,

however, for FS-specific business practices. The additional training will cost $2,000 per day for two weeks. This includes a training consultant to run classes at FS headquarters. With travel and lodging, the total cost is $23,000.

- *Ongoing consulting:* Once the system is up and running, FS will need to pay for consultants to help maintain the system. FS estimates that if it budgets for consultants to come in once a month for $3,000 a day, it should be able to have all employees' questions answered. The total yearly cost of additional consulting is $36,000.
- *Computer maintenance:* FS needs to make sure all PCs and servers run properly. To do this, FS would purchase a maintenance contract to cover all hardware. This contract costs about $1,000 a month, or $12,000 yearly.
- *Network and database administrator:* FS would need a full-time network and database administrator to run the system. Salary, including benefits, for a skilled person is $200,000 per year.

Option 2: Using ERP Through an Application Service Provider

The other option is to use an ASP to deliver ERP software. Estimated costs for this option are as follows:

- *PCs:* FS still estimates that it must purchase 10 new PCs because many more users will now be accessing the computer system. Each PC costs $1,500, for a total of $15,000.
- *Maintenance on PCs:* The maintenance contract on all the PCs at FS costs $600 per month, or $7,200 yearly.
- *Software through the ASP:* The monthly cost of delivering ERP software to FS over the Web is $33,333, or $400,000 yearly.
- *Training:* Training of FS employees is provided by the ASP as part of the monthly software fee.

Make a recommendation to FS. Which option should FS choose—purchase the SAP software and computers outright, or use an ASP? Set up a spreadsheet that will add all the costs of each option. In each scenario, you must deal with the net present value (NPV) of money.

NPV is a way to figure out whether an investment is profitable, or in this case, to compare outlay of funds from one method to another. NPV can be calculated over a number of years; in our case, we need a five-year outlay of funds for the ERP project. The syntax of NPV is =*NPV (hurdle rate percentage, range of values)* in an Excel spreadsheet. The values in the range can be positive or negative numbers. In our case, they are all outflows, but we can work with them as positive numbers. The hurdle rate is the rate of discount over the period. The hurdle rate is the minimum acceptable rate of return on a project that a company will accept.

Your spreadsheet should begin like the one shown in Figure 8-3 (with years continuing through 2012). Assume payments are made throughout the year, as opposed to lump sums at the beginning of the year.

ERP Purchasing Options							
Option 1 - Buying computers and software outright							
Items				**2008**	**2009**	**2010**	**2011**	**2012**
Database server				70000				
Application server				40000				
10 PCs				15000				
Software - SAP				500000				
Consultants - initial (6 months)				486000				
Training (2 weeks)				23000				
Consultants - maintenance (1 day per month)					36000	36000	36000	36000
PC maintenance					12000	12000	12000	12000
Network administrator				200000	200000	200000	200000	200000
Total				1334000	248000	248000	248000	248000
NPV				$1,646,671.81				
Option 2 - Using an ASP								
PCs				15000				
PC maintenance					7200	7200	7200	7200
ASP cost				400000	400000	400000	400000	400000
Total				415000	407200	407200	407200	407200
NPV				$1,224,277.26				
Hurdle rate				20%				

FIGURE 8-3   Cost comparisons: buying versus renting

To complete this exercise, perform the steps that follow:

1. Calculate the cost of the two methods of implementing SAP ERP for five years. Use the spreadsheet illustrated in Figure 8-3 as your guide. Use the NPV calculations to reference the hurdle rate at the bottom of the spreadsheet. Vary the hurdle rate, following the directions your instructor provides.
2. Consider using different hurdle rates for each option. Why might varying hurdle rates be applicable for this decision?
3. Write a memo, with your spreadsheet attached, to the CIO. Answer this question: Which method should FS choose, and why? Be sure to consider both the qualitative aspects and the quantitative aspects of the choice. Also address the viability of the ASP.

Your instructor might assign the following additional exercises:

1. Use the Internet or library resources to research the use of ASPs. Find cases of companies that have been successful in using an ERP system through an ASP. Describe one success story in a memo to your instructor.
2. Think about whether you would give different advice to a smaller company than you would to a medium-sized company or to a large company, regarding the use of ASPs. Why?

NETWEAVER

E-commerce is driving companies to connect their business applications, such as ERP, to the Internet to provide data sharing between companies. The combination of software tools that lets various programs within an organization communicate with other applications is called **Web services**. Whereas an application service provider (ASP) delivers software over the Web, Web services connect various software applications over the Web. Not surprisingly, ERP companies are trying to stake out their territory in this new and lucrative field. SAP has invested a lot of time, money, and energy into its Web services platform, NetWeaver. NetWeaver, like other Web services products, allows various vendor applications to share data.

Companies are warming to the idea of Web services, also known as SOA, or service-oriented architecture. *Information Age*'s Effective IT research report has found that 50 percent of enterprises have some sort of SOA strategy. One benefit of adopting SOA is the ability to add new applications, quickly making the organization more responsive. SOA also relies on open standards, allowing easier integration of software and offering the potential to reuse computer code, which would reduce the time and cost of implementing new systems. This aspect of SOA is certainly enticing, compared with traditional systems that are often cumbersome and time-consuming to implement. However, implementing SOA is not easy. The IT analyst group Ovum reports that one in five U.S. companies implementing SOA have experienced "unexpected complexity."

The return on an SOA investment is often difficult to determine. According to a study published by Nucleus Research, only 37 percent of 106 organizations polled claimed that their SOA projects had a positive ROI. Respondents indicated that the main benefit of SOA was the ability to reuse computer code.

NetWeaver Tools and Capabilities

SAP's NetWeaver is a collection of components that support business transactions over the Internet. Included are modules named Enterprise Portal, Mobile Infrastructure, Business Intelligence, Master Data Management, and Exchange Infrastructure.

The Enterprise Portal also goes by the name mySAP.com. It gives users complete access, or a portal, to all their work on a single screen, using links to all major aspects of their jobs. A **portal** is a customizable Web site that serves as a home base from which users navigate the Web. The Enterprise Portal acts as a central access point to a company's intranet, operating through a secure link on a browser. For example, a user in the Finance department could set up Enterprise Portal with links to SAP ERP financial transactions, as well as links to financial metrics for the company, stock market indices, e-mail, a calendar, and other information important for that person's job.

The advantage of having a personalized portal is its efficiency. A user only has to log on to one system to get all the information needed to perform a job. Without a portal, users often have to log on to multiple systems, such as an ERP system, industry exchanges, or suppliers' sites. Transferring information between systems is frequently difficult. With the Enterprise Portal, all information is available through the Web services provided by NetWeaver. All the important links are presented in one screen, and transferring data is simplified by the ability to "drag-and-relate" data from one area to another.

Mobile Infrastructure is another module of NetWeaver. It allows users to access and work with data through mobile devices such as PDAs, cell phones, and pagers. Mobile Infrastructure provides remote access to data within the SAP system and other data contained within a company's information system. The benefits are obvious. A salesperson could use her PDA to see a customer's historical order information while in the middle of a sales visit at the customer's office. Connecting SAP and VoiceObjects AG adds voice capability to NetWeaver, allowing users to enter data to the SAP CRM system using voice commands from their cell phones. Linking the SAP and VoiceObjects AG systems can be done without middleware or any changes to the software.

Another facet of NetWeaver is Business Intelligence (BI), which incorporates a data warehouse and data mining tools. BI can be delivered in a personalized manner with Enterprise Portal. It integrates information from various sources and processes, both within and outside of the firm. BI works with any database management software and any operating system that is running NetWeaver. Datamonitor predicts that the market for business intelligence software will double by 2012, reaching $8 billion in revenue. In 2007, SAP acquired Business Objects, and Oracle acquired Hyperion; the bought-out firms are both providers of BI software.

ANOTHER LOOK

BI at Rohm and Haas

Rohm and Haas, a specialty chemicals company, employs over 16,000 people in 100 different locations in 27 different countries. The company expanded rapidly from 1998 to 2006, almost doubling in size through 45 acquisitions and divestitures. At that point, the company had 300 different IT systems and was generating 600 different reports. The company sorely needed Business Intelligence to get the reporting situation under control.

Rohm and Haas had begun an SAP implementation. The company developed a data warehouse that now generates BI reports. Previously, employees downloaded data into a spreadsheet to produce reports for upper management. Now employees prepare the same reports using dashboards, consolidated interfaces providing quick access to various software components. Employees access the dashboards through NetWeaver.

Rohm and Haas developed key performance indicators (KPIs) to incorporate within the dashboards. KPIs measure the state of the organization. For example, a gross profit KPI is one indicator of success. With the new dashboard system, a manager looking at the gross profit KPI can drill down to further levels of detail behind problems and opportunities.

Rohm and Haas is using three dashboards: a financial-based KPI dashboard for executives; a sales, standard gross profit, and volume KPI-oriented dashboard called the Pulse, which is for a wider range of users; and the Reporting and Analysis Toolkit. The Pulse is the most heavily used of the three dashboards because of its ability to compare daily sales to the month-to-date sales and compare both daily and month-to-date sales to budgeted sales. Users can also view these data broken down by business. Figure 8-4 shows some of the reports available through the executive dashboard.

continued

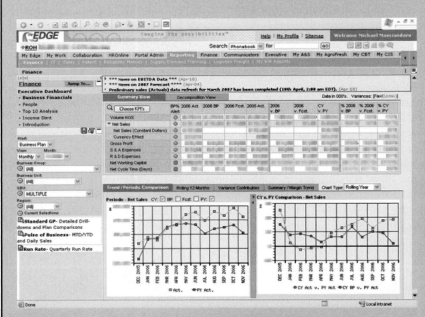

FIGURE 8-4 Rohm and Haas executive dashboard

The success of the system is in its usage. The dashboard system is used by 4,100 employees, including Rohm and Hass CEO and president Raj Gupta, who uses it every day. The dashboards are allowing employees to be more proactive rather than reactive in their decision making.

Question:

1. Choose an industry that interests you, such as financial services. What are the KPIs for a company in that industry? How can those KPIs be measured?

Another NetWeaver feature, Master Data Management, provides data consistency within a company's SAP system. For example, at Fitter Snacker, the two sales groups, Direct and Wholesale, might have had different numbering systems for common customers. Master Data Management would ensure that the customer numbers are the same. The grocery industry could save $25 billion to $50 billion if suppliers could synchronize their data, such as product numbers, with retail outlets. NetWeaver allows this seamless Web interface to ensure proper data synchronization.

NetWeaver's Exchange Infrastructure allows different applications to share data. By adhering to the standard of the Exchange Infrastructure, companies don't have to write code to enable different applications to transmit data. For example, using Exchange Infrastructure, a business can keep its current EDI system and seamlessly integrate that with its ERP system. SAP's Web Application Server, the development environment that is the foundation of NetWeaver, gives Exchange Infrastructure its customizability.

Many nonusers of NetWeaver are confused as to what it is. SAP admits that the concept of tying applications together is not easy to explain in a few sentences in a marketing brochure. In response to this confusion, SAP has sponsored the *SAP NetWeaver for Dummies* publication and is also providing information sessions for its customers around the world.

NetWeaver at Work for Fitter Snacker

Now we will examine how NetWeaver could help Fitter Snacker. Assume that the two top salespeople, Amy Sanchez and Donald Brown, are busy selling NRG bars directly to customers and to distributors. Amy works from home. She logs on to the SAP system with her laptop computer, using the SAP GUI. She doesn't know much about the SAP system, nor does she have to. She needs to know how to place customer orders and check on their status. When Amy goes on a sales call, she brings her notepad and calculator with her to jot down orders and quotes. When she returns home, she plugs those numbers into the SAP system and confirms her quotes. Amy would like to have some additional information on how salespeople in other regions are doing and what mix of bars they are promoting, but she doesn't know how to access any of that information. She also would like to see if there are new ways to market to her customers.

Donald Brown is also a salesperson, but he deals with distributors. He has been chosen to be a tester for the new NetWeaver SAP server. Every day, Donald comes into the office and logs on to his Enterprise Portal, which was tailored for his job. He sees figures from his top 10 customers, data on production and inventory of bars, the current stock quote for Fitter Snacker on the NASDAQ exchange, the current market price for oats, wheat germ, and honey, his e-mail, and the local weather report. Today, Donald will make an important sales call for a regional grocery chain. He grabs his wireless PDA and some extra business cards, and heads out the door. During lunch with the purchasing agent for the grocery chain, Donald is able to check up-to-the-minute details on current sales orders and can confirm promises to ship additional bars next week, thanks to SAP's Mobile Infrastructure. Back at the office, Donald calls up the Business Intelligence module in NetWeaver. From there, he can run a few reports to find out which snack bars are currently selling better nationwide, grouped by region and time of year. He can also analyze snack bar sales using data mining, to find sales patterns that can help him plan future sales calls.

Exercise 8.3

1. After reading about the features of SAP's NetWeaver and looking at how Amy Sanchez and Donald Brown perform their duties at Fitter Snacker, try to convince Fitter Snacker's CEO to implement NetWeaver. Write a memo to the CEO outlining your arguments.

2. The CEO is impressed with your work but has asked you for an ROI (return on investment) analysis. How do you begin doing that? What numbers do you need, and who are the people you would have to interview to get those numbers?

DUET

Microsoft and SAP have been working on a software product, Duet, intended to let companies access SAP data and processes using the familiar Microsoft Office interface. The goal of the project is to expand and simplify the adoption of SAP ERP by making workers more efficient. While the Duet product has numerous advantages, it also brings its own challenges. Companies must be using a relatively current version of SAP ERP, and must run Microsoft server software as well. Some Duet features require the company to be using other SAP products, such as NetWeaver and CRM.

A challenge for the Duet product is the growing competition between SAP and Microsoft in the ERP software market. Microsoft is moving its ERP software product, Microsoft Dynamics, to large companies while SAP continues to develop software products aimed at smaller companies. This situation, in which the two software companies are cooperating in some areas and competing fiercely in others, has been termed "co-opetition." Some analysts believe that the competitive-cooperative relationship will benefit customers by make it easier for more users to access ERP data.

ACCESSING ERP SYSTEMS OVER THE INTERNET

In addition to the standard GUI that is part of every ERP system, ERP vendors now offer access to their systems through a Web browser such as Microsoft Internet Explorer or Netscape Navigator. Users and systems administrators find it much more efficient to access their ERP systems through the browser; this method avoids the time-consuming installation of the GUI. This method of accessing ERP systems usually appeals to a company's IT staff as well, since the company does not need to distribute the GUI software to hundreds or thousands of users.

The University of Delaware's new PeopleSoft HR system was originally accessed through a PeopleSoft GUI. Since the University of Delaware processes most of its HR requests using forms on the Web, PeopleSoft agreed to create access to the system through a Web browser. Users can log on to the university's PeopleSoft system from home or from a remote office, as long as they have a computer with an Internet connection and Web browser. They do not have to be in their own offices to perform their jobs. System administrators at the university also find that software upgrades are much simpler to manage. Only the software on the server needs to be upgraded—there is no special software installed on user PCs.

XML

Extensible Markup Language, shortened to **XML**, is the new programming language of the Internet. XML uses tags that define the data contained within them. Similar to data types assigned to records in a database, XML tags apply specific meaning to the data within a Web page. XML-coded data can go directly from a Web page into a database without having to pass through middleware or, worse yet, be rekeyed into the system. In comparison, most Internet pages are written in Hypertext Markup Language, or HTML. HTML specifies only how your information will look (by assigning text styles, coloring, placement of graphics, and so on) when viewed through a browser.

XML users can create their own tags. Many companies are working together to create industry-specific tags to use in conducting e-commerce with each other. Fitter Snacker could keep its raw material records in XML format for easier transfer from suppliers' systems. An example of an FS XML document is shown in Figure 8-5. The customized tags in the document describe, or define, the data much as a database would. The ingredients honey and oats are defined as *descriptions* of an *item*.

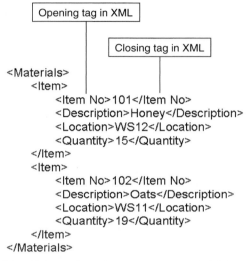

```
Opening tag in XML

                Closing tag in XML

<Materials>
    <Item>
        <Item No>101</Item No>
        <Description>Honey</Description>
        <Location>WS12</Location>
        <Quantity>15</Quantity>
    </Item>
    <Item>
        <Item No>102</Item No>
        <Description>Oats</Description>
        <Location>WS11</Location>
        <Quantity>19</Quantity>
    </Item>
</Materials>
```

FIGURE 8-5   Fitter Snacker document in XML

Another example of the benefits of XML can be found in the insurance industry. Often in insurance, data must be passed between different documents. Sometimes the documents are from different companies. With XML, data are tagged and flow into those disparate documents without any additional processing. Every person involved in the chain of events—insurance brokers, insurance companies, underwriters, and insurance agents—can streamline their paper processing with XML.

ERP systems are now ready to accept data in XML format. Using XML, companies can transfer data from their Web sites (data such as order information) directly into their ERP systems. This approach streamlines data entry, reduces errors, and reduces server loads.

XML is also very attractive to smaller companies. Small companies often transfer data over telephone lines or using fax machines. Using XML makes electronic data transfer much more affordable. Without XML, a company might receive text-formatted data over the Internet. The data need to be converted from the text format to a database record—a potentially time-consuming and expensive process. Using XML tags to define the data would eliminate the conversion requirement.

ANOTHER LOOK

XML, ERP, and E-Commerce

Occidental Chemical Corporation, a subsidiary of Occidental Petroleum, uses SAP for its chemicals and resins business. When OxyChem linked its SAP system with a customer's SAP system, both companies realized tremendous customer service benefits. Oxy-Chem has since embarked on a mission to link its SAP system with all of its customers. OxyChem has 5,000 customers of varying size in different industries. The company formed an internal team, which included customers, to study the project. The strategy they developed was to offer four different methods to integrate the systems.

The first method offers an XML-based ERP system-to-ERP system linkage between OxyChem and the customer. Once the two systems are linked, OxyChem can check the customer's ERP system to find out when it is becoming low on various raw materials. Purchase orders are automatically generated and orders flow efficiently.

The second method uses physical probes inside the customer's raw material containers. Using a Web interface and sales forecasts, the system automatically sends a purchase order in XML directly into OxyChem's SAP system.

In the third method, customers go to the ChemConnect chemical trading site (which you learned about earlier in this chapter) to order chemicals from OxyChem. These orders are created in XML, so they can flow directly into the SAP system at OxyChem.

The last, and most popular, method is an OxyChem-owned Web portal that allows customers to place orders directly with OxyChem. The company originally thought this portal would be used only by its smaller customers, but customers of all sizes use it.

Of course, OxyChem still maintains a call center for those customers that do not want to order by computer.

Questions:

1. Why did OxyChem give customers a choice in how to integrate their systems?
2. List the benefits of linking a manufacturer's ERP system with a customer's ERP system.
3. Draw a process map (see Chapter 7 for details on process mapping) to depict the flow of information from a manufacturing plant using a chemical from OxyChem. Show the process of requesting and then filling the order.

RADIO FREQUENCY IDENTIFICATION TECHNOLOGY

Radio frequency identification technology, known commonly as **RFID**, is becoming an efficient way of tracking items through a supply chain. An RFID device is a small package, or tag, that includes a microprocessor and an antenna, and can be attached to products. The location of an item with an RFID tag can be determined using an RFID reader, which emits radio waves and receives signals back from the tag. The reader is also sometimes called an interrogator because it "interrogates" the tag. Since microprocessors continue to become more powerful and less expensive, according to Moore's Law (discussed in Chapter 2), RFID technology has reached a point where it is inexpensive enough to be cost-effective. Today, most materials are tracked using bar codes and bar-code readers. Bar-code labels can degrade in bad weather, and an employee has to point a bar-code reader at the bar code to read it. RFID technology does not need this line-of-sight connection, and can withstand most environmental stresses.

Wal-Mart is in the process of implementing an RFID system for its supply chain. Suppliers are shipping pallets of goods to Wal-Mart marked with RFIDs. Wal-Mart reads the RFID tags on cases and pallets when inventory enters a stockroom, when those cases of pallets go to the floor, and, ultimately, when empty cases are discarded. Much of the data collected during RFID reads is passed on to Retail Link, Wal-Mart's Web-based software that lets its thousands of buyers and suppliers check inventory, sales, and more. In 2007, Wal-Mart was still in the midst of bringing its suppliers on board with the technology, which was taking longer than anticipated, with only 600 of the 20,000 suppliers using the RFIDs by the end of the year. Wal-Mart shifted its focus to the use of RFIDs in the stores, rather than in the distribution centers. Goods coming from the warehouse have RFIDs and are scanned upon arrival at the store; the tags are not used for items on the shelves. Wal-Mart employees also scan tags on discarded boxes to determine stock levels. The suppliers to Wal-Mart who have adopted the RFID technology do not see any return on investment except keeping the Wal-Mart account. Wal-Mart does admit that the RFID project has not met its goals. In fact, some industry observers speculate that Wal-Mart became distracted with the RFID initiative, losing focus on new products and relationships with customers.

Pharmaceutical firms are also working toward adopting RFID technology, to comply with upcoming FDA regulations that would require track-and-trace technology on all drug packages to prevent counterfeiting. Johnson & Johnson uses RFID technology to keep track of heart stents. Stents have a three-month shelf life, so managing inventory is time-sensitive, but the stents need to be available at the hospital for emergency heart surgeries. Since each stent is worth about $2,500, it pays for the company to keep track of every stent. Johnson & Johnson can read the data remotely and deliver stents to hospitals that are in need.

Companies involved in manufacturing and selling RFID technology, including IBM, claim that their customers' projects experience a return on the RFID investment in less than 18 months. Consumer goods companies, with their large numbers of customers, are especially interested in applying this new technology, so they can track products better as they move through the supply chain. The British retail giant Marks & Spencer has been tagging items in one pilot store. To avoid privacy concerns (such as tracking the wearer of a garment), the RFIDs are easily detachable. Eddie Dodd, the chief technology officer of British Telecom's Auto-ID services, which provides the RFID readers to Marks & Spencer, says, "RFID undoubtedly gives you clear visibility into what's happening to your supply chain with a degree of accuracy." Another U.K. giant, the food chain Tesco, piloted RFIDs with the goal of integrating the technology on the individual item level, but recently scaled back its RFID plans. Tesco is now placing RFIDs on roll cages (which hold items such as milk) and pallets.

Procter & Gamble is using RFID technology to collect information about the sales of its products. The company uses its RFID-gathered data on fluctuations in sales of Pampers and Luvs diapers to help manage its supply chain processes. Although P&G sells disposable diapers to a population that is fairly constant, a slight change in the number of diapers sold at stores can be amplified by traditional supply chain planning processes, causing large fluctuations in demand throughout the supply chain, a phenomenon known as the **bullwhip effect**. There are two primary reasons that the bullwhip effect occurs. First, the relatively smooth daily demand for products gets aggregated into occasional larger orders from stores to distribution centers. The stores and distributors then aggregate these large orders into even larger and less frequent orders to the manufacturer. The second source of demand fluctuation is the human behavior that results from the potential for a delay in receiving a product. Like people waiting for an elevator who repeatedly push the call button in an attempt to speed the elevator's arrival, companies sometimes place multiple orders for a product when they suspect a potential shortage or delay, only to cancel the additional orders when the deliveries begin to arrive. The bullwhip effect has been understood for a long time. The only way to alleviate this problem is to have better information for planning, which is what RFID technology promises to provide. To figure out how many unsold diapers are in a particular store, someone has to either count all the packages on the shelf and in the back room, or use a bar-code reader and scan them. RFID tags eliminate the manual counting process. Determining the number of items in inventory in a building becomes automatic. With RFID, real-time inventory data are always available. RFID tags store detailed data, such as production lot number and production date, which provides even more benefit when items have an expiration date or are subject to recall, like Fitter Snacker's line of NRG bars.

SAP's ERP software is RFID-ready. Along with partner company Infineon Technologies AG, SAP provides tools to easily link RFIDs to back-office systems. Through NetWeaver, SAP can integrate RFID data into both SAP and non-SAP applications.

Chapter Summary

- E-commerce is transforming the way companies do business. Business-to-consumer e-commerce can streamline a company's ordering operations and record information about customers that can be used to customize sales and promotional activities, making the company more competitive.

- Business-to-business e-commerce is changing the way companies buy and sell goods. New forms of procurement such as auctions, reverse auctions, and trading exchanges—all with dynamic pricing—are replacing the traditional intermediary.

- ERP is an essential component for all forms of e-commerce. An integrated information system is required to provide speed and consistency in transaction processing and other back-office operations.

- Application service providers (ASPs) are allowing companies to use ERP without a large initial investment, making ERP systems available to smaller companies. There are risks associated with using an ASP, however, and the decision to buy or lease must be weighed carefully.

- Web services, or service-oriented architecture, offers a combination of software tools that lets various programs within an organization communicate with other applications.

- SAP's Web services platform is NetWeaver, which includes tools for seamless connectivity of diverse applications through the World Wide Web. NetWeaver also includes modules such as Business Intelligence, Mobile Infrastructure, and Master Data Management.

- Users of ERP systems often access those systems through a Web browser, rather than the ERP systems' graphical user interface (GUI). Using a Web browser rather than a GUI program simplifies the software maintenance task.

- XML, Extensible Markup Language, defines data on a Web page. ERP systems are using XML to integrate systems between suppliers and customers for easy data transfer.

- RFID devices, or radio frequency identification devices, are used in tracking items in transit. RFIDs are particularly useful in supply chain processes for shipping and receiving cases and pallets of items. ERP vendors are developing the capability to incorporate RFID technology into ERP software.

Key Terms

Application service provider (ASP)

Back-office processing

Bullwhip effect

Business-to-business (B2B)

Business-to-consumer (B2C)

Commodity

Denial of service (DoS)

Electronic data interchange (EDI)

Electronic marketplace

Exchange

Extensible Markup Language (XML)

Internet-based procurement

Portal

Radio frequency identification (RFID)

Reverse auction

Value-added network (VAN)

Web services

Exercises

1. Assume you have just graduated and you land a job at Fitter Snacker as its new purchasing agent. The person you replaced has just retired after 35 years in the job. You are eager to put to use some of the skills you learned at school, especially the work you did on B2B e-commerce. Write a memo to your new boss, the vice president of Supply Chain Management, on the virtues of e-commerce and why Fitter Snacker should now do Internet procurement.

2. Define application service provider. Do some research on the Web about current opinions of using ASPs. Try to find some examples of success and failure stories. Under what circumstances do you think using an ASP makes sense?

3. Go to the library and check out several publications about NetWeaver. Choose one aspect of NetWeaver and write a paper describing its functionality.

4. RFID technology is changing rapidly. Assume you have landed a summer job in the production department at Fitter Snacker. One of your jobs is to brief the entire team on the status of RFIDs. Create a PowerPoint presentation that brings the FS crew up to speed on this emerging technology.

5. Using the library or Internet sources, write an update on the status of FDA legislation surrounding RFID packaging on drugs. Are there any privacy concerns?

For Further Study and Research

Angwin, Julia. "Top Online Chemical Exchange Is an Unlikely Success Story." *The Wall Street Journal,* January 8, 2004.

Barrett, Larry. "Who Will Win the SAP, Oracle Battle?" ASPnews.com, August 28, 2007. http://www.aspnews.com/trends/article.php/3696666.

Bradshaw, Tim. "RFID suits M&S." *Information Age*, June 19, 2006. http://www.information-age. com/briefing_room/old_briefing_rooms2/information_management/implementation/rfid_ suits_marks.

eMarketer.com. "Europe E-Commerce to Grow Fourfold from 2003 to 2006." July 2, 2003.

"ERP+ERP=B2B." *Integrated Solutions,* July 2002.

Fox, Pimm. "Private exchanges drive B2B success." *Computerworld,* May 7, 2001. http://www.itworld.com/Tech/3478/CWD010507STO60197.

Franke, Jon. "Microsoft's latest ERP push: What does it mean for SAP, Duet?" SearchSAP.com, March 21, 2007.

Gaskin, James E. "XML Comes of Age." *InternetWeek.com*, April 3, 2000.

Herm, Marcus. "XML - An opportunity for small and medium-sized enterprises." XML - The Site, http://www.softwareag.com/xml/library/herm.htm.

Kollmann, Tobias. "Measuring the Acceptance of Electronic Marketplaces: A Study Based on a Used-car Trading Site." *Journal of Computer-Mediated Communication 6,* no. 2 (January 2001). http://jcmc.indiana.edu/vol6/issue2/kollmann.html.

Konicki, Steve, and Rick Whiting. "Let's Keep This Private." *Information Week,* July 30, 2001.

Maxcer, Chris. "Rohm and Haas: Dashboards to the Rescue." *SAP NetWeaver Magazine*, Fall 2007.

McDougall, Paul. "Closing the Last Supply Gap." *Information Week*, November 8, 2004.

Mooraj, Hussain. "Pharma RFID Adoption - Retail All Over Again?" RFID Update.com, May 8, 2006. http://www.rfidupdate.com/articles/index.php?id=1112.

SAP Solution Brief: Exchange Infrastructure. http://www.sap.com/platform/netweaver/pdf/BWP_SB_ExchangeInfrastructure.pdf

SAP Solution Brief: SAP Master Data Management. http://www.sap.com/search/index.epx?q1=Solution%20Brief%3A%20SAP%20Master%20Data%20Management

Sullivan, Laurie. "Fast Track to Success." *Information Week,* June 21, 2004.

———. "IBM Takes RFID Services to Midsize Companies." *Information Week,* September 20, 2004.

———. "Wal-Mart's Way." *Information Week,* September 27, 2004.

Surowiecki, James. "EZ Does It." *The New Yorker,* September 8, 2003.

Swabey, Pete. "Most SOA projects bring no ROI." *Information Age*, September 6, 2007.

———. "Structural hazard." *Information Age*, March 19, 2007.

Vamosi, Robert. "Botnets for sale." CNET.com, March 23, 2007. http://reviews.cnet.com/4520-3513_7-6719515-1.html?tag=feat.2.

"VoiceObjects Technology to be Integrated in SAP NetWeaver Phone Application Server Technology to Enable Voice-Driven Telephone Access to SAP Applications." VoiceObjects.com, Press Release, March 13, 2007. http://www.voiceobjects.com/en/news/2007/031307.html.

Waigum, Thomas "How Wal-Mart Lost Its Technology Edge." *CIO*, October 4, 2007.

Weier, Mary Hayes. "Wal-Mart Rethinks RFID." *Information Week*, March 26, 2007.

Westervelt, Robert. "Duet: SAP customers see success, challenges ahead." SearchSAP.com, August 8, 2006.

Williams, Kathy. "How Secure Is E-Commerce?" *Strategic Finance* (March 2000): 23.

Woods, Dan, and Jeffrey Word. *SAP NetWeaver for Dummies.* Indianapolis, Indiana: Wiley Publishing, Inc., 2004.

Accounting and Finance (A/F) A functional area of business that is responsible for recording data about transactions, including sales, raw material purchases, payroll, and receipt of cash from customers.

Accounts receivable Accounting information that records money a customer owes for the goods received.

Activity-based costing An advanced form of inventory cost accounting in which overhead costs are assigned to products, based on the manufacturing activities that gave rise to the costs.

Advanced Business Application Programming (ABAP) SAP ERP internal programming language.

Application service provider (ASP) A business that delivers software applications to companies over a network. Sometimes that network is the Internet.

Archive Permanently stored data.

Asset Management (AM) module A module in SAP ERP that helps a company to manage fixed-asset purchases (plant and machinery) and related depreciation.

Audit trail Linked set of document numbers related to an order.

Back office processing The processing of sales orders through a company's marketing, manufacturing, shipping, and accounting systems.

Balance sheet This summary of a company's account balances includes cash held, amounts owed to the company by customers, the cost of inventory on hand to be sold, long-term assets such as buildings, amounts owed to vendors, amounts owed to creditors, and amounts that the owners have invested in the company.

Best practices The application of the best, most efficient ways in which business processes should be handled.

Bill of material (BOM) The "recipe" listing the materials (including quantities) needed to make a product.

Bullwhip effect Large fluctuations in demand throughout the supply chain caused by a slight change in the number of products sold.

Business function Business activities within a functional area of operation.

Business process A collection of activities that takes one or more kinds of input and creates an output that is of value to a customer. Creating the output might involve activities from different functional areas.

Business process innovation (BPI) The process of improving processes.

Business process reengineering A quality improvement philosophy that recommends radical change to achieve radical improvements.

Business-to-business (B2B) Communication and sales between manufacturers, wholesalers, retailers, and suppliers. This communication can occur using both EDI and the Internet.

Business-to-consumer (B2C) Communication and sales between businesses and the buying public. Popularly, but incorrectly, thought of as the most common form of e-commerce.

Capacity The amount of an item that can be produced.

Cash-to-cash cycle time   The time from paying suppliers for raw materials to collecting cash from the customer (used in supply-chain management metrics).

Client-server architecture   Data stored in a central computer (a server) are downloaded to a local PC (a client of the server) where data are processed. Historically, client-server architecture replaced many companies' mainframe-based architecture.

Commodity   An item that is widely available at a standard level of quality.

Condition technique   A SAP control mechanism that accommodates the various ways that companies offer price discounts.

Continuous improvement   A quality improvement philosophy that prescribes systematic and repeated improvement efforts.

Controlling (CO) module   A module in SAP ERP that is used for internal management purposes. The software assigns a company's manufacturing costs to products and to cost centers, facilitating cost analysis.

Cost variance   The difference between actual costs and standard costs.

Currency translation   Converting financial-statement account balances expressed in one currency into balances expressed in another currency.

Customer master data   Central database tables in SAP ERP that store permanent data about each customer. Master data are used by many SAP ERP modules.

Customer relationship management (CRM) software   A variety of different software tools that use data from a company's ERP system to enhance the company's relationships with its customers. CRM software allows these activities: segmenting customers, one-to-one marketing, sales-force automation (SFA), sales-campaign management, marketing encyclopedias, and call-center automation.

Data mining   The statistical and logical analysis of large sets of transaction data, looking for patterns that can aid decision making and improve customer sales and customer service. Data mining is often done with data in a data warehouse.

Data warehouse   A database, separate from a company's operational database, that contains subsets of data from the company's ERP system. Users analyze and manipulate data in the warehouse. Thus, they do not interfere with the workings of the database that is used to record the company's transactions.

Database management system (DBMS)   The technology that stores database records in an organized fashion and allows easy retrieval of the data.

Delivery   In SAP, release of the documents that a warehouse uses to pick, pack, and ship orders.

Denial-of-service (DOS) attack   A security attack on a Web site or Web server in which attackers block access to the Web-based service through a variety of means.

Deployment flowcharting (swimlane flowcharting)   A type of flowchart that depicts team members across the top, with each process step aligned vertically under the employee or team working on it.

Development (DEV) system   In a SAP system landscape, one of three separate SAP systems; DEV is used to develop configuration settings for the system using ABAP code.

Direct costs   Costs in a finished product that can be estimated fairly accurately.

Document flow   The linked set of document numbers related to an order; an "audit trail."

Drill down   The ability to view the details behind a summary of information.

Dynamic process modeling   A method of evaluating process changes before they are implemented by putting into motion a basic process flowchart using computer simulation techniques to facilitate the evaluation of proposed process changes.

Electronic commerce (e-commerce)   The buying and selling of goods and services over the Internet.

Electronic data interchange (EDI)   A computer-to-computer transfer of standard business documents that allows companies to handle the purchasing process electronically, avoiding the cost and delays resulting from paper-based systems.

Electronic marketplace   A gathering place for buyers and sellers on the Internet.

Enterprise Resource Planning (ERP)   ERP systems help to manage business processes such as marketing, production, purchasing, and accounting in an integrated way. ERP does this by recording all transactions in a common database that is used by information systems throughout the company and by providing shared management-reporting tools.

Error log   A record of discrepancies that occur during a payroll run.

Event process chain (EPC)   A graphic model of a business process that uses only two symbols: events and functions.

Exchange   A type of B2B electronic marketplace that focuses on a single industry.

Extensible markup language (XML)   An Internet programming language that uses tags that define the data contained within them.

Financial accounting   The documenting of all transactions of a company that have an impact on the financial state of the firm. The documented transactions form the basis for reports, or financial statements, for external parties and agencies.

Financial Accounting (FI) module   A module in SAP ERP that records transactions in the general ledger accounts and generates financial statements for external reporting purposes.

Flowchart   Any graphical representation of the movement or flow of concrete or abstract items.

Functional areas of operation   A categorization of business activities, including marketing, sales, production, and accounting.

Gap analysis   An assessment of disparities between an organization's current situation and its long-term goals.

General ledger   A traditional record of accounting.

Hierarchical modeling   The ability to flexibly describe a business process in greater or less detail depending on the task at hand.

Human capital management (HCM)   Another term for Human Resources that describe the tasks associated with managing a company's workforce.

Human resources (HR)   A functional area of business that manages recruiting, training, evaluation, and compensation of employees.

Human Resources (HR) module   A module in SAP ERP that facilitates employee recruiting, hiring, training, and payroll and benefits processing.

Income statement   A financial statement that shows a company's profit or loss for a period of time.

Indirect costs   Overhead items that are difficult to associate with a specific product.

Information system (IS)   The computers, people, procedures, and software that store, organize, and deliver information.

Initial fill rate   The percentage of the order that the supplier provided in the first shipment to the manufacturer or retailer (used in supply-chain management metrics).

Initial order lead time   The time needed for the supplier to fill an order (used in supply-chain management metrics).

Integrated information system   An information system that allows sharing of common data throughout an organization. ERP systems are integrated systems because all operational data are located in a central database, where they can be accessed by users throughout an organization.

Intercompany transaction   A transaction that occurs between a company and its subsidiary.

Internet-based procurement   The use of Internet technologies for procurement activities.

Job   A generic description of an employee's work responsibilities or the position that the person holds.

Lead-time   The cumulative time required for a supplier to receive and process an order, take the material out of stock, package it, load it on a truck, and deliver it to the manufacturer.

Legacy system   An IS that is already in existence prior to installation of an ERP system.

Lot sizing   The process for determining purchase and production order quantities.

Managerial accounting   Accounting that deals with determining the costs and profitability of the company's activities.

Marketing and sales (M/S)   The functional area of business that is responsible for developing products, determining pricing, promoting products to customers, and taking customers' orders.

Master production schedule (MPS)   The production plan for finished goods.

Materials Management (MM)   A module in SAP ERP that manages the acquisition of raw materials from suppliers (purchasing) and the subsequent handling and storage of raw materials, work in process, and finished goods.

Material Master Data   Central database tables in SAP ERP that store relatively permanent data about materials. These data are used by SD, MM, and other SAP ERP modules.

Materials Requirements Planning (MRP)   A production-scheduling methodology that determines the timing and quantity of

production and purchase-order releases to meet a master production schedule. This process uses the bill of material, lot-size data, and material lead-times.

Metrics   Measurements of performance; discussed in this book in relation to the effects of supply-chain management efforts.

Modules   Individual programs that can be purchased, installed, and run separately, but all extract data from the common database.

MRP Record   The standard way of showing the Manufacturing Requirements Planning process on paper.

On-demand   The practice of CRM software and computer equipment residing with the CRM provider.

On-time performance   A measure of how often a supplier meets agreed-upon delivery dates (used in supply-chain management metrics).

Open architecture   Software that allows integration with third-party software. SAP ERP is an example of open-architecture software. The term can also be applied to hardware products.

Organizational change management   The supervision of human behavior aspects of organizational change.

Organizational structure   The method used in SAP ERP to define the relationships between organizational groups such as companies, plants, storage locations, sales divisions, and distribution channels.

Overhead   A company's cost of operations, such as the costs for factory utilities, general factory labor, factory management, storage, insurance, and other manufacturing-related activities. Overhead is often called an indirect cost of production.

Payroll run   The process of determining each employee's pay.

Person   The unique individual who holds a position.

Plant Maintenance (PM) module A module in SAP ERP that allows planning for preventive maintenance of plant machinery and managing maintenance resources, so equipment breakdowns are minimized.

Portal A customizable Web site that serves as a home base from which users navigate the Web.

Position A job that might exist in more than one department. For example, the administration assistant position might be required in the Wholesale Department as well as in Inside Sales.

Process boundary In process mapping, a definition of those activities to be included in the process, and what is considered part of the environment.

Process mapping A type of flowcharting that specifically represent pictorially the activities occurring within an existing business process.

Product cost variant The procedure for developing a product cost analysis.

Production Planning (PP) module A module in SAP ERP that maintains production information; production is planned and scheduled, and actual production activities are recorded.

Production (PROD) system In a SAP system landscape, one of three separate SAP systems; the actual system that the company uses to run its business processes.

Profit and loss (P&L) statement A record that shows a company's sales, cost of sales, and the profit or loss for a period of time.

Project System (PS) module A module in SAP ERP that allows planning for and control over new R&D, construction, and marketing projects. This module allows for costs to be collected against a project budget, and it can be used to manage the implementation of ERP itself.

Qualifications Skills or abilities associated with a specific employee.

Quality circles A quality improvement technique in which employees in a department have regular team meetings to discuss problems and collaboratively develop solutions.

Quality assurance (QAS) system In a SAP system landscape, one of three separate SAP systems; the system where testing is done.

Quality Management (QM) module A module in SAP ERP that helps to plan and record quality-control activities, such as product inspections and material certifications.

R/3 The first integrated information system released by German software vendor SAP in 1992; now called SAP ERP. This ERP system contains the following main modules, which can be implemented as a group or selectively: Sales and Distribution (SD), Materials Management (MM), Production Planning (PP), Quality Management (QM), Plant Maintenance (PM), Human Resources (HR), Financial Accounting (FI), Controlling (CO), Asset Management (AM), Project System (PS), Workflow (WF), and Industry Solutions (IS).

Radio frequency identification (RFID) A tracking technology that uses a small package, or tag, device that includes a microprocessor and antenna that can be attached to products. The location of an item with an RFID tag can be determined using an RFID reader, which emits radio waves and receives signals back from the tag.

Raw data Data on sales, manufacturing, and other operations that have not been analyzed.

Remuneration elements The part of an employee's pay include the base pay, bonuses, gratuities, overtime, sick pay and vacation allowances that the employee has earned during the pay period.

Repetitive manufacturing A manufacturing environment in which production lines are switched from one product to another similar product.

Requirements Skills or abilities associated with a position.

Return on investment (ROI)   A ratio calculated by dividing the value of the project's benefits by the value of the project's cost.

Reverse auction   An auction that features one buyer and many sellers.

Rough cut planning   A common term in manufacturing for aggregate planning.

Sales and Distribution (SD) module   A module in SAP ERP that records sales orders and scheduled deliveries.

Safety stock   Extra raw material and packaging kept available to help avoid stockouts.

Sales forecast   A company's estimate of future product demand, which is the amount of a product that customers will want to buy.

SAP ERP   ERP software produced by SAP; previous versions were known as R/3 and mySAP ERP.

Scalability   Information systems are deemed "scalable" if their capacity can be extended by adding servers to the network, rather than replacing the entire system. Scalability is a characteristic of client-server networks, but usually not of mainframe-based systems.

Scope creep   The unplanned expansion of a project's goals and objectives, causing the project to go over time and over budget, as well as increasing the risk of an unsuccessful implementation.

Service-oriented architecture (SOA)   Software that enables systems to exchange data without complicated software links, also called Web services.

Short list   Up to three top candidates for a position, each of whom will be interviewed.

Silo   An unintegrated information system configuration in which individual business functional areas each have their own hardware, software, and methods of processing data and information.

Software modules   See "modules."

Standard cost   The expected cost of manufacturing a product during a particular period. Standard costs for a product are established by (1) studying historical cost patterns in a company and (2) taking into account the effects of current manufacturing changes.

Statutory and voluntary deductions   Paycheck withholdings that include taxes (federal, state, local, Social Security, and Medicare), company loans, and benefit contributions.

Stockout   A manufacturing shortfall that occurs when raw materials or packaging run out.

Succession planning   Outlining of the strategy for replacing key employees when they leave the company.

Supply chain   All of the activities that occur between the growing or mining of raw materials and the appearance of finished products on the store shelf.

Supply Chain Management (SCM)   Sharing long-range production schedules between a manufacturer and its suppliers, so raw materials can be ordered and delivered in a timely manner, thus avoiding stock-outs or excess inventory.

Tasks   The assigned responsibilities related to a specific job.

Tolerance group   Ranges that define limits on the dollar value of business transactions that an employee can process.

Transport directory   A special data file location on the DEV server that stores changes to the SAP system landscape.

Value added   An increase in a product's or service's value, from a customer's perspective.

Value-added network (VAN)   An EDI service provider. Companies acquire EDI service by subscribing to a VAN's EDI network.

Value analysis   Analysis of each activity in a process for determining the value the activity adds to the product or service.

Web services   A combination of software tools that lets various programs within an organization communicate with other applications.

Workflow (WF)   A set of tools in SAP ERP that can be used to automate any of the activities in SAP ERP. It can perform task-flow analysis and then prompt employees (by e-mail) if they need to take action.

Workflow tasks   In SAP ERP, links between work and various transactions. These links can include basic information, notes, and documents, as well as direct links to business transactions.

INDEX

S